Native Space

Native Space

Geographic Strategies to Unsettle
Settler Colonialism

Natchee Blu Barnd

FIRST PEOPLES
New Directions in Indigenous Studies

OREGON STATE UNIVERSITY PRESS   CORVALLIS

Library of Congress Cataloging-in-Publication Data

Names: Barnd, Natchee Blu, author.

Title: Native space : geographic strategies to unsettle settler colonialism /
    Natchee Blu Barnd.

Other titles: Geographic strategies to unsettle settler colonialism

Description: Corvallis : Oregon State University Press, 2017. | Series: First
    peoples: new directions in Indigenous studies series | Includes bibliographical
    references and index.

Identifiers: LCCN 2017028002 | ISBN 9780870719028 (original trade pbk. :
    alk. paper)

Subjects: LCSH: United States—Civilization—Indian influences. | Indians of
    North America—Ethnic identity. | Indians of North America—Land tenure.
    | Street names—United States. | Names, Geographical—United States. |
    Whites—Relations with Indians. | Indians of North America—Colonization.
    | Decolonization—United States. | Indian art—United States. | Human
    geography—United States.

Classification: LCC E77 .B22 2017 | DDC 973.04/97—dc23

LC record available at https://lccn.loc.gov/2017028002

♾ This paper meets the requirements of ANSI/NISO Z39.48-1992
(Permanence of Paper).

First published in 2017 by Oregon State University Press
Second printing 2018
Printed in the United States of America

Oregon State University Press
121 The Valley Library
Corvallis OR 97331-4501
541-737-3166 • fax 541-737-3170
www.osupress.oregonstate.edu

Contents

Illustrations

Acknowledgments

I want to offer some brief words of thanks to all who helped bring this project to reality. I start with my many colleagues for their encouragement, inspiration, mentorship, and friendship. At Oregon State University, Juan Herrera and Bradley Boovy read my manuscript in various forms and deserve special attention for helping it reach final form, and not enough can be said of their efforts. Bradley also contributed a much needed photo! Ron Mize and Juan Antonio Trujillo graciously offered up spaces in their homes for me to use as writing retreat spaces, and provided the crucial time and distance needed to get my writing completed. A wide variety of support came from my Ethnic Studies faculty colleagues Patti Sakurai, Marta Maldonado, Daniel López-Cevallos, and Robert Thompson. Hours of support and fun came from our office manager Leonora Rianda. Likewise from the staff of the School of Language, Culture, and Society, Loretta Wardrip and Karen Mills. Amanda Green provided many valuable research hours while a graduate student, helped me hone portions of this project, and also obtained a needed photo. Big thanks to Kenny and Carrie Maes for watching my children enough times to help us feel sane and thus allow me to continue my work.

Other research assistance came from generous librarians, scholars, and staff from across the country, including: Margaret Bender at Wake Forest University, Julie Monroe in Special Collections and Archives at the University of Idaho Library, staff at the Hood Museum at Dartmouth University, the Kansas State Historical Society, and Deborah Carroll from Oregon State University Valley Library Inter-Library Loan.

The Center for the Humanities at Oregon State University granted much-needed release time and a welcome, collegial atmosphere when I needed to change directions and generate new lines of research. Big thanks to Angie

Morrill for introducing me to Mary Braun and recommending my work. My eternal thanks to Mary Braun at OSU Press for then recruiting me, and to the First Peoples Initiative and the Mellon Foundation for providing a publishing forum for new scholars such as myself. The now defunct summer write-up grant for the School of Language, Culture, and Society provided needed time away to complete several chapters. Thanks all the baristas in JavaStop of the OSU Memorial Union, who kept me fueled, kept me company, and kept me laughing.

My gratefulness to the anonymous reviewers at the *American Indian Culture and Research Journal* for thoughts on the article that became chapter two, and to those reviewing my manuscript for Oregon State University Press. I send appreciation to several of my UC San Diego mentors, who continue to offer their support and guidance long after my break from student status: Paul Spickard is an invaluable role model and sounding board at a moment's notice, Ross Frank who is always my supporter and even worked as my research assistant for the manuscript (check out the photo credits!), and Yến Lê Espiritu who models a balanced and thoughtful scholar. I have a special place in my heart for Edwina Welch for holding it down at UC San Diego's Cross-Cultural Center for so long and so well, and thus helping me feel confident that at least one place in the world had my (and others') best interests in mind. Out of that center came many lifelong friendships/families, and thus my gratitude to Jason Perez and Denise Pacheco for listening to and giving productive and honest feedback on my mock job talk, which later became chapter three and helped me land my first academic job.

I want to thank a number of friends and colleagues in the various Native communities I discuss in this book or have worked with in my teaching. This includes Kathy Cole, Bobby Mercier, and David Harrelson at the Confederated tribes of Grand Ronde; Buddy Lane at the Confederated Tribes of Siletz; Mina Seminole of the Northern Cheyenne at Lame Deer; and Renissa Walker, Garfield Long, Mollie Grant, and David Wachacha in the offices of the Eastern Band Cherokee. My unlimited appreciation to Vanessa Campbell at the Musqueam Language and Culture Department, as well as the Musqueam elders and community members who read chapter one and offered their guidance on Musqueam community history and language. I am indebted to Herman Fillmore, who works in cultural resources and language instruction with the Washoe tribe, and was extraordinarily helpful in tracking down a number of leads and helping me better understand the Dresslerville community. He also arranged for his sister,

Helen Fillmore, to take photos for me, some of which are included in the book! Thanks to Helen for her great eye and wonderful photographic contributions.

Certainly not least, I send all my love and affection to my life partner Charlene, and our amazing and powerful daughters. You each made this project possible and inspired me to always keep in mind what was most important about the research and my responsibilities to others.

Introduction

Indigenous geographies proclaim "we are still here" in a most grounded way. In the context of a settler colonial world, they serve as reminder of presence despite centuries of material, philosophical, and social structures founded on producing Native absence. Indigenous continuations also illustrate that geographies are not simply places. Choices, ways of understanding the world, and actions create spaces that exist in particular ways. These choices, understandings, and actions, then, must be continually practiced and reaffirmed in order for any given space to continue to exist. Indigenous geographies have quietly overlapped and coexisted in tension with the geographies of the settler colonial state. They have been submerged, but not eliminated. While they have changed to survive the violences directed at eliminating this overlap and coexistence, indigenous peoples have sustained Native spaces.

One of the contributions this book makes is to demonstrate how Native people are making those spaces. I am particularly interested in the everyday or mundane ways this happens. As I elaborate below, I think of the mundane in two ways. On the one hand, my attention to everyday spatial activities refers to seemingly noncontroversial practices and policies in tribal communities that nevertheless help confirm and sometimes redefine indigenous geographies. On the other hand, I am also interested in intentionally controversial or evocative artistic practices employed by Native artists who in their work must still rely on and build their "insurgent messages" based on frameworks available from everyday indigenous spatialities (Slocum 2007).

Given the distinctive space-centered epistemologies that form the bases of indigenous relationships with the land it is not surprising that indigenous communities and individuals seek to re-narrate place or reclaim indigenous geographies rather than merely capitulate to the force of national "progress" and inclusion. While attending to postcolonial spatial tension, this book centers Native space-making practices in order to illustrate how indigenous geographies persist within and confront the US settler colonial nation. I proceed under the understanding that indigeneity and space are mutually bound frameworks, and yet they are in need of attention given the urgent context of settler colonialism. By focusing on the fundamental relationship between indigeneity and space, I contend that we can better recognize the decolonizing possibilities and actualities of indigenous geographies. We can also gain an understanding of how indigenous geographies operate as crucial acts of self-determination and cultural continuity. Finally, we can see how Native spaces serve as analytics for rejecting Native dispossession and the interlocking logic of incorporation within the multicultural nation-state.

A brief outline of the book is provided at end of this introduction. For the rest of this opening, however, I turn to some definitions and frameworks that guide this interdisciplinary project. The following concepts pull from a vast array of theoretical and methodological influences, including comparative ethnic studies, critical toponymies, indigenous geography, ethnohistory, performance theory, language and translation theory, postcolonial theory, Whiteness studies, cultural studies, cultural geography, history, American Indian studies, social history, and critical cartography. I start with the concepts indigeneity, Indianness, and inhabiting. One of the additional goals of this book it to highlight the spatiality of Indianness and indigeneity. By examining diverse articulations and interconnected uses of Indianness in the construction of indigenous and settler spatialities, *Native Space* addresses the relationship between race, space, indigeneity, Whiteness, and colonialism in the contemporary United States. I use a number of terms throughout this book whose meanings are interconnected and therefore can benefit from elaboration. In the discussion that follows, I treat several as clusters, including *colonialism*, *settler colonialism*, *postcolonialism*, and *neocolonialism*, followed by *spatiality*, *indigenous spatialities*, and *settler spatialities*.

Indigeneity

Indigeneity, or what might be loosely defined as the "quality of being indigenous," is deeply embedded within and defined by colonial contestations over land and geography (Radcliffe 2017,1). If settler colonialism is fundamentally defined by its spatial organization and outcomes, then so too must be indigeneity, a term and concept that codes as the supposed precondition of, as well as ongoing foil to, colonial completion. Indigeneity originates in and relies on colonial interventions and acts of racialized differentiation, yet also overlaps with self-definitions from those whose ancestors were present on the continent before European arrivals.

If invoking indigeneity always signals the process of contested land claims and occupations of North American lands, then the term can only make sense through some basic relational understandings of presence, belonging, and history. It tells us who was here first, who came later, and who should remain. It locates fundamental cultural differences and positions them as either rooted in practices developed in relation to this specific landscape, or else developed elsewhere. It tells us how the environment came to be upon the moment of colonial contact, and what happened afterward. It frames the meaning of states and nations, who decides those meanings, and what implications follow.

Although the idea of the indigenous is dependent on and created through colonial encounter, I also emphasize that both Native and non-Native geographies must deploy divergent frames of indigeneity. When indigenous peoples are ignored, invisibilized, marginalized, or mythologized—all of which are standard practice in the United States—these acts reproduce fundamental frameworks for the European colonization and ongoing US American occupation of North American lands. At the same time, the settler colonial nation and its citizens often invoke indigeneity in order to inhabit moral and geographic authority, usually through a co-optation of the "Indian." In contrast, Native peoples invoke indigeneity to mark belonging and relationships to this land as well as to contest colonization and the White possessive (Moreton-Robinson 2015). These contestations over indigeneity matter because they either deny or prepare us for after-colonial geographies, or the spaces of possibility that can emerge should we attend to settler colonialism and critiques of White supremacy.

The promise of the extension of indigenous geographies posits an effort toward transforming human relationships with the world in such a way as to recover nonexploitative engagements and to restart the responsibilities of settlers and arrivants toward indigenous peoples and cultures. This is clearly a revolutionary and structurally radical imagining, and cannot be accomplished merely by changing the faces of those in control of a racially hierarchical, capitalist, and colonial geography. Yet, these geographies already coexist in uneven fashion, whether conceived as rival, differential, or simply indigenous geographies (Castree 2004; Ferguson 1985; Goeman 2013; Ingold 2007; Said 1993; Stark 2012).

Before we can more fully entertain these imaginative but already partially actualized possibilities, we must more thoroughly assess how indigenous and settler colonial geographies persist. To work toward this task, this book attends to the practice of *inhabiting* as one of the powerfully mundane or "common sense" ways spaces are enacted, justified, and sustained. I forefront inhabiting in order to clarify the spatiality embedded within indigeneity and Indianness, and to both highlight and distinguish between the kinds of everyday (spatial) practices that produce either settler or indigenous geographies (Billig 1995; Rifkin 2013, 2014).

Indianness and Inhabiting

Indigeneity requires some engagement with the related concept of the Indian, a racial construct that has long facilitated the dispossession, subjugation, and attempted incorporation of Native peoples into the United States (Barker 2005, 16–17; Byrd 2011, xxiii). I intentionally use Indianness in my analysis, in addition to indigeneity, because I want to actively frame how the processes of creating indigenous space in a settler colonial nation must simultaneously attend to the tensions of overlapping and often opposing geographies. Indianness encompasses a dialectical and sometimes oppositional set of understandings about Native peoples in what is now the United States. Indianness references indigenous self-definitions as well as definitions that are externally imposed and sometimes mythological. It refers simultaneously to the supposedly self-evident identity category of "Indians" as well as all the varied meanings generated within and across diverse and complicated Native communities and histories. Certainly the concept of an Indian has also come to serve as a useful shorthand for individuals' grounded personal or tribal experiences, for pan-tribal identifications, and for acts of strategic essentialism (Hertzberg 1981; Spivak 1987). In these ways, "Indianness" is used by but not fully owned by Native peoples (Berkhofer 1978). In fact, the "Indian" is usually an abstract, even fictional conjuring: a fundamental but ultimately nominal figure in the US national trajectory. Ignoring for the moment the relative simplicity and artificial cleanliness of this binary (Native/non-Native), we can see that Indianness contains a tension that is continually being negotiated and stretched into service by different constituencies (Bird 1996; Deloria 1998; Green and Massachusetts Arts and Humanities Foundation 1975). Sometimes these meanings overlap or reinforce one another. In other contexts, they clash and battle. Thus, I intend the term "Indianness" to signal both non-Native usages of the Indian toward the production of space and the Native dis/engagements with those appropriative and imposed usages toward the same purpose.

While I draw attention to Indianness as a fluid and multiply constituted symbol, it is important to note that I do not intend to imply that Native peoples are fully contained by this dialectic process. Contestation over these definitions and meanings is certainly a core element of what it means to be indigenous. Part of my intervention, however, is to consider precisely the ways that Native individuals and communities have always expressed and generated new and self-determined notions of identity, culture, and sovereignty that are not necessarily just rooted in a response to the violently narrow notions of Indianness imposed from the outside (Carpio 2004; Goeman 2013; Simpson 2014). For these reasons, I am attentive to the dialectic of the Indian, although in most instances I will deliberately privilege the more self-proscribed and self-determining practices of indigeneity and indigenous space-making that are less concerned with directly contesting appropriations and land claims. Thus, I have gathered a series of examples to show how indigenous geographies also emerge from relatively self-contained efforts firmly rooted in and ultimately constitutive of Native-centered worlds. I argue that indigenous geographies can never be just a response to settler colonialism if they signal the continuation (however adaptive or appropriative) of precolonial epistemologies, ontologies, and practices. To think otherwise is to assume completion of the colonial project, to freeze history and space, and thus to encapsulate and ventriloquize indigeneity solely via Eurocentric and state logics.

This brings us to the concept of inhabiting. At its base, inhabiting signals the moment(s) when a body is situated in a particular physical location. It is also a verb, implying some sort of spatially defined and relational set of actions. Inhabiting describes a frame used for establishing belonging or home, a relation to place. In this book, inhabiting is sometimes rooted in possession, both of land and of Indianness. Thus, this term also refers to spatial production: to the process of making meaning in relation to the land where bodies are situated. I apply the term "inhabiting" to signal differing notions of relationships to land (broadly defined to include air, water, underground, and so on) and the related processes of legitimization for bodily presence in specific locations (whether individual or collective).

In terms of indigenous geographies, for example, we can turn to Tim Ingold's wonderful discussion of inhabiting applying only when describing humans being fully "immersed in the fluxes of the medium [air], in the incessant movements of wind and weather" (Ingold 2007, S34). From this perspective, humans fundamentally inhabit the air, not the land. Native space likewise tends to be based on inhabitants that "make their way *through* a world-in-formation," intimately accounting for and centering the processes and relations between elements like land, rock, water, air, clouds, smoke, wind, and weather (Ingold 2007, S32; emphasis original). This contrasts with modern Eurocentric models that position humans as "exhabiting" the surface of the planet, and thus being "stranded on a closed surface" and seeing the world only through metaphors of interior or contained spaces. Such a frame explains the desire and impetus to extend control and shape the nonhuman world and to ignore processes and relationships except where directly harvestable. This illustrates a core difference between indigenous and settler geographies. This conception of inhabiting as a frame of reference for engagement with context can also reframe settler colonial engagement with indigeneity.

Within the settler colonial frame, inhabiting points to the European legal construct that delineates a discrete and static moment in time that forever renders European presence legitimate. This reconciliation of belonging continually emerges through cultural constructs that rely on layered and symbolic inhabitations beyond the legal repertoires of occupation and must be performed repeatedly to address the "refractory imprint of the native counter-claim" (Wolfe 2006, 389). Indeed, if settler colonialism is a structure rather than event, as Patrick Wolfe suggests, then I argue that inhabiting Indianness represents one of the necessary modes for ongoing settling and the process of sustaining settler geographies (Wolfe 2006, 388).

As a mode of presence, justification, and relation, indigenous persistence and indigenous geographies also require inhabiting. When Native peoples re-inhabit Indianness (or manifest indigeneity), they signal the ongoing ways that indigenous peoples reject White possession of the Americas that appears so inevitable and yet invisible to non-Native peoples. "For indigenous people," Aileen Moreton-Robinson notes, "white possession is not unmarked, unnamed, or invisible; it is hypervisible" (Moreton-Robinson 2015, xiii). Thus, in the face of the overwhelming possessive logics of Whiteness and settler colonialism, indigenous peoples sustain Native geographies that unsettle and create "ontological disturbance." The various remappings and assertions of space practiced by indigenous peoples discussed in the following chapters therefore have been selected because they either intentionally confront or analytically provide counterpoint to what Mark Rifkin identifies as "settler common sense." Rifkin describes settler common sense as rooted in colonial policy and practice but emerging through and consolidating via everyday and affective experiences that allow settlers' "access to Indigenous territories . . . to be lived as given, as simply the unmarked, generic conditions of possibility for occupancy, association, history, and personhood" (Rifkin 2013, 323).

My use of "inhabiting" thus builds on Moreton-Robinson's notion of a "white possessive" that racially frames settler colonialism and on Rifkin's attention to the crucial and reproductive role everyday enactments (or productions) of space play in materializing and sustaining the logics of possession and the formal mechanisms of dispossession (Rifkin 2013, 337). Perhaps most usefully, inhabiting reminds us how spatial enactments can be practiced and (re)arranged in sometimes unexpected ways toward different kinds of relations to lands, or different geographies.

Colonialism, Settler Colonialism, Postcolonialism, and Neocolonialism

At its simplest, "colonialism" stands for the "conquest and control of other people's lands and goods," although we must be aware that the modes and configurations of such "conquest and control" vary greatly (Loomba 2015, 20). Much of the variation among colonialism and its derivative forms and practices centers on difference in the modes and methods of control. In terms of the actual mechanisms for control, we must consider whether people/labor, goods/resources, or land serve as the primary vehicle for conquest, although they can and usually do overlap in meaningful ways.

As a project concerned with the production of space (which I describe below), interrogating and comparing these various modes and methods can help us better understand how current formations of power continue to hinge on colonial-era land claims and conflicting geographies, and how they vary and shift. For most of this text, I return to the two concepts that seem most appropriate: settler colonialism and neocolonialism. These terms (and attention to these practices) can also help keep questions of land and space forefront. As Patrick Wolfe notes, "territoriality is settler colonialism's specific, irreducible element" (Wolfe 2006, 388). Land, of course, also remains core to indigenous identities, histories, and cultures (Deloria 1994; Deloria and Wildcat 2001; Pierotti and Wildcat 2002). Thus, I intend my use of these terms as reminders that colonizing projects initiated during European travels across the globe continue to manifest tensions over land and space, including in some mundane ways. As these concepts remain an ever-present context, much of this book therefore considers the various ways that indigenous peoples call attention to and resist the ongoing nature of colonial space-making, as well as the ways they continue to maintain and produce their own geographies against or despite colonialism.

Colonialism, in its "purest" form, has traditionally been understood as an unequal power relationship wherein a dominating population extracts labor and/or resources from a subordinated population of an "external" location. "A colonial system of relationships," Lorenzo Veracini points out, "is premised on the presence and subjugation of exploitable 'Others'" (Veracini 2014, 615). This basic frame also highlights the embedded role of geography as a central defining factor of all colonial endeavors. In short, colonialism fundamentally describes a geographic relationship, one in which "differing" geographies serve as a mechanism for producing and maintaining unequal power relations with a "home" geography.

The seventeenth- and eighteenth-century sugar plantations established by Europeans in the Caribbean represent one concrete example of colonialism: in this case, plantation colonialism. The history of these plantations illuminates a unique element in European colonialism. European colonialism developed mutually with concepts of race and the corresponding practices of racism, as well as with capitalism (Blaut 1993; Césaire 1972; Goldberg 1993). In the Caribbean, lands had already been largely widowed of indigenous peoples, and thus plantation labor was extracted by enslaving and importing Africans. These ventures figured into a global economic expansion, as European nations vied against one another for economic and political superiority through exploitation of lands and laboring bodies. As part of European global exploration, developing notions of race both shaped and were shaped by the slave trade and the colonial encounter with non-Europeans. Through the frame of a newly emerging racial logic targeting non-Europeans, the European "idea of the colonial world became one of a people intrinsically inferior, not just outside history and civilization, but genetically pre-determined to inferiority. Their subjections was not just a matter of profit and convenience but also could be constructed as a natural state" (Ashcroft, Griffiths, and Tiffin 1998, 47).

The scale of European colonialism, starting in the sixteenth century, extended beyond that of previous examples of colonies and conquests. Increased mobility and the codevelopment of European colonialism with capitalism greatly increased the geographic reach of such projects and fortified each through expanding and increasingly interdependent economic networks. As Ania Loomba suggests, "colonialism was the midwife that assisted at the birth of European capitalism, . . . without colonial expansion the transition to capitalism could not have taken place in Europe" (Loomba 2015, 22).

Settler colonialism describes a form of colonialism wherein nonindigenous or "settler" populations implant themselves in new lands. Lorenzo Veracini describes the difference as being between shaping and controlling a landscape versus changing and claiming the landscape. "In the case of colonialism what is reproduced is an *(unequal) relationship*, while in the case of settler colonialism, what is reproduced is a *biopolitical* entity" (Veracini 2014, 627; emphasis original). Patrick Wolfe summarizes the same process by offering this succinct assessment: "Settler colonialism destroys to replace" (Wolfe 2006, 388). Settlers initiate a fundamental transformation in the demographics, cultures, and physical landscape of colonized lands. Settler presence is the core feature of this mode of domination, with the goal of establishing a new home to solidify territorial claims.

As an extension of the home country, then, settler colony lands are redesigned toward the home country's imposed goals, as well as toward reflecting settler identity. As Said notes, "Colonial space must be transformed sufficiently so as no longer to appear foreign to the imperial eye" (Said 1993, 226). This transformation can, of course, take numerous forms. For the purpose of this book, I focus tightly on the model characterizing British North American settler colonialism, in which indigenous peoples were not broadly incorporated into the settler societies, and in which they usually resisted such efforts when pursued. Disease often devastated indigenous populations, yet settlers also actively used warfare as well as political, "legal," and other-than-legal means to actively dispossess them of lands.

This articulation, of course, constructs a simplified binary between those who settle and those who are already present, and streamlines the complex and uneven process by which these outcomes unfold. In terms of the settlers, for example, Jodi Byrd notes that settler colonialism recruits people from both the colonizing nation and beyond in the form of servants, laborers, slaves, immigrants, and refugees (Byrd 2011). She thereby differentiates between settlers and what she calls "arrivants," those largely non-White, nonindigenous peoples that likewise arrive and occupy the land even as they do not arrive under the same circumstances or positionalities. Most important for this book, however, is to recall that the categories of indigenous, settler, and arrivant peoples always rest on the relationship between physical presence on a specific land and belonging, on relationships to home and belonging. Those categories point to ongoing and conflicting *practices* of space-making (inhabiting) resulting from colonial processes and the complex modes of presence and proscribed inclusion.

In this context we must note how the dominant settler colonial geographies continually work to submerge indigenous ones. As Ania Loomba explains, for settler colonialism to exist, "the process of 'forming a community' in the new land necessarily meant un-forming or re-forming the communities that existed there already, and involved a wide range of practices including trade, settlement, plunder, negotiation, warfare, genocide, and enslavement" (Loomba 2015, 20). Likewise Paige Raibmon and Cole Harris have outlined the various structured/official and everyday/unofficial means by which settler colonial societies have to make and "unmake" indigenous space precisely because Native peoples work to recover, maintain, or reinscribe their geographies (Harris 2002; Raibmon 2008).

I also want to emphasize that the relationship between colonialism and settler colonialism often proves more complementary and less binary. Veracini, for example, points out how, historically, colonialism and settler colonialism might best be seen as a "division of colonial labour" (Veracini 2014, 627), whereby these two different models of conquest flexibly operate at various scales, and historically can be found to cooperate in extending, protecting, or securing mechanisms and structures of domination. Recognizing this relationship is important for making sense overall of the processes and fluidities of domination as well as the varied forms of resistance required to address the ongoing consequences of mutable colonialisms. This recognition is reflected in the terms "neocolonialism" and "postcolonial," which I outline below. Those terms emerge from understanding how colonialism is continually reshaped and resisted. They serve as reminder that any effective contestation must also evolve.

Veracini argues that finding appropriate analogies for colonialism and settler colonialism can tell us a great deal about how we might contest such forms of domination. In his "heuristic" analysis, he creatively analogizes colonialism and settler colonialism with virus and bacteria, respectively. The virus operates in largely parasitic fashion, sustaining itself through the lifeblood of the host and mutating to best ensure its survival and success through various hosts and host defenses. Historically, human colonies likewise extract resources and compel colonized labor for sustenance, while maintaining their relatively distinct coherence as entities. Bacteria, in contrast, "attach to surfaces and form aggregations" that reproduce without direct exploitation of a host, and in the process of aggregation take on new forms as entities (Veracini 2014, 615). Bacterial colonies effectively absorb, assimilate, and transform their environment such that they "make and remake places and are also simultaneously transformed by them" (Veracini 2014, 624). In short, such colonies adapt to a new environment, rapidly reproduce and expand, and finally stabilize as a new and unique entity. A settler colony parallels bacteria in that its vitality relies on the mutual transformations of the colonizing "body" and the space of colonization, even as the indigenous population may not strategically figure into those transformational processes.

Reminding us that these processes can operate in concert, Veracini helps point out historical conditions in which "some areas could only become subjected to colonizing metropoles after colonial 'viruses' had evolved in ways that would allow it to *penetrate* as well as to *attach* to new areas" (Veracini 2014, 619; emphasis original). Thus, a settler colony sometimes requires the groundwork of a preceding colony. Here we should note that, despite the differing and sometimes cooperating methods for exogenous prosperity, both methods share and are premised on spatial domination (what Veracini calls "destination locales"). In settler colonialism, however, land rather than people proves the most immediate mechanism for domination and the core point of contention for both colonizer and colonized.

I draw on these analogies because they are useful not only in thinking about contestation of colonial structures and outcomes, but also in extending our understanding of the relationships of colonialism and settler colonialism to that of postcolonialism and neocolonialism. As Blunt and McEwan explain, "the 'post' of 'postcolonialism' has two meanings, referring to a temporal aftermath—a period of time *after* colonialism—and a critical aftermath—cultures, discourses and critiques that lie *beyond*, but remain closely influenced by, colonialism" (Blunt and McEwan 2002, 3; emphasis original). The notion of postcolonialism as an analytical tool and as an account of the "critical aftermath" of colonialism is therefore closely tied to a recognition of the emergence of neocolonialism, which encompasses global economic domination regardless of historic colonial relations and carves out new forms of domination operating entirely through the forces of globalization.

This project uses the concept of postcolonialism to focus on those critical intellectual and material interventions against colonial and *ongoing* neocolonial practices still in need of confrontation. I am interested in those indigenous-centered postcolonialisms that seek to disrupt the ongoing experiences of settler colonialism and neocolonialism. Neocolonialism, which directly translates to a "new" form of colonialism, sustains persistent structures of "cultural, economic, and political inequalities" and perpetuates the "endurance of colonial discourses" that originate with colonization and yet "persist long after the end of formal political colonization" (Nash 2002, 220). Ashcroft and colleagues point out that Kwame Nkrumah (who first coined this term in his book *Neo-Colonialism: The Last Stage of Imperialism*) identified the ways neocolonialism was actually "more insidious and more difficult to detect and resist than the older overt colonialism" (Ashcroft et al. 1998, 163; Nkrumah 1966). This articulation of what Veracini labels a "mutated" and more evasive "strain" of colonial domination points to the ways that neocolonialism operates through hegemony rather than through direct force, and thus the mechanisms of domination and the resulting spatial configurations are more easily naturalized and less easily confronted (Gramsci 1971; Veracini 2014).

Often still overlooked is the way that settler colonialism continues, and that it heavily overlaps with neocolonialism. This overlap points directly to the role of geography. English (and now multicultural) settler societies currently present a confluence of direct intervention and indirect neocolonial structures in relation to indigenous peoples. When the United States is positioned as a postcolonial nation or is denied as an example of domination in relation to indigenous peoples, it furthers the entrenchment still reliant on a land base predicated on continual indigenous dispossession.

Spatiality, Indigenous Spatialities, and Settler Spatialities

When I use the terms "spatiality" and "spatialities," I am actively marking and recalling the fact that space is a production, and is always multiple. In *The Production of Space,* Henri Lefebvre provides the core theoretical reframing of space as something other than a simple blank stage upon which social actors gather and interact (Lefebvre 1991). He explains that space is a product of our social imaginings and actions, which coalesce into coherence as well as material form. Spatial productions express and secure dominance most effectively when seen as merely existing—as supposed natural expressions of the world-as-it-is. In this way, spatiality signals the individual and collective processes we engage in to produce space and the ways that we are also produced by spaces.

One of the important benefits of a conception of space as produced and contingent is that geography can then be more fully understood in relation to power. Any dominant form of space or spatiality stands as, and is, power, as it structures particular values about, views of, and practices within the world and reinforces these structures by shaping encounters to match that world. Thus, an analysis of space must fundamentally hold the exercise of power as one of its principal features. As John Allen argues, "power is *inherently* spatial and, conversely, spatiality is *imbued* with power" (Allen 2003, 3; emphasis original). In the context of settler colonialism and neocolonialism, we can readily see how space is imbued with power since it is not only hegemonic in conveying a sense of the geography of the nation-state as being just "common sense," but it has also been actively utilized in dispossession and disempowerment toward the benefit of one group of peoples over another. In the aftermath of the various forms of colonialism, the dominant contemporary geographies still represent a successful consolidation and extension of the forms of spatial production initiated centuries ago.

While current understandings of space already imply a "relational" or processual practice, combining spatiality with a modifier—indigenous, settler—further signals two specific kinds of space and space-making operating in tension with one another (Massey 2005). The additions of the qualifiers "settler" and "indigenous" then leads to the task of clarifying or qualifying what defines these different kinds of spatiality. I am drawing on the categories of indigenous and settler spatialities to note two shapes of engagement centered on the relationship to the lands of North America and the frames for making sense of those relationships. Most importantly, these different spatialities are rooted in the historic and racialized experiences of peoples who experienced colonialism as either colonized or colonizer. In this sense, these categories of spatiality can be viewed in relation to one another, intimately formed by the experience of encounter and subsequent reconfigurations of land, culture, and agency. At the onset of colonialism, the Doctrine of Discovery predicated conferral of dominion on both the inability and the unwillingness of Europeans to recognize or respect indigenous spatialities (treaty or not). Thus, we find concepts like terra nullius and the "virgin landscape," both of which relied on a Western spatiality rooted in intentional, observable, and demarcated human interventions in the processes of the natural world. This world was thus quickly overlaid with abstract space to render it recognizable, manageable, and alienable. While indigenous peoples obviously modified the land, such labor and engagement was not always signposted, and their modifications worked effectively enough within the existing ecosystems as to often remain invisible to Western eyes (Anderson 2013; Cajete 2000; Cronon 1983; Kimmerer 2013). Settlers simply interpreted indigeneity as either lacking proper spatiality or without sufficient authority and moral capacity. The result was conscription of the land into settler spatial systems that erased "other ways to relate geography and identity" (Radcliffe 2011, 140).

Such conflicts continue. These frames remain the central differences between indigenous and nonindigenous peoples, and they continue to coexist in uneasy tension with one another in the same place and time. "Spatial co-habitation by Indigenous and settler populations" Brad Coombes and colleagues note, "entails confrontation of divergent notions of place construction, along with other disorderly ontological categories which underpin epistemological and teleological classifications" (Coombes et al. 2011, 486).

Organization of the book

Before briefly summarizing the chapters that follow, I want to note a couple of methodological approaches and highlight the relationship between the first three chapters and the final two. Each chapter begins with a narrative that shares some of my own experiences with specific spatial productions and how they exemplify spatial practices. The narratives themselves represent important methodological practices in a couple of ways. I am consciously emphasizing the power and function of the mundane, or what Michael Billig calls the "banal," everyday acts that prove crucial for dominant spatial productions (Billig 1995). Billig's study of everyday acts that sustain nationalism explains how seemingly meaningless and nonspectacular activities actually represent core sets of artifacts and regular repertoires through which nationalism is understood and on which it is dependent. These small acts effectively set the stage on which "larger" and more explicit enactments of nationalism can take place and make sense. Victoria Freeman similarly observes such acts in the erasures of indigenous Toronto. Following Alan Gordon, she reminds us that formal public memory events and collective performances of ideologies must be successfully crafted in advance "to have any symbolic or emotive power" during obvious and ceremonial performances of identity and ideology (Freeman 2010, 30; Gordon 2001, 165).

These small, usually mundane, acts are therefore crucial as the ongoing labor required for any spatial production. Such practices are not limited to hegemonic productions. Although indigenous geographies, for example, must be fluent in the dominant spatial regimes and practices as a matter of survival (and result from assimilation violences), they can be sustained and produced only through normalizing practices. Native space must be constantly recognized and made viable through daily practices. In this way, my approach further illustrates the everyday spatial work being done through what Mark Rifkin has nicely delineated as "settler common sense," as well as frames the impact of practicing embodied rather than just legal or political forms of indigenous sovereignty (Bruyneel 2007; Coulthard 2014; Lyons 2010; Rifkin 2014; Warrior 1995). My movement across the chapters is therefore intended as a loose progression from the more concrete forms of space-making toward the more conceptual and artistic, which is also a movement from less to more explicit and self-conscious spatial counterproductions. Viewing the chapters together, the specific examples reveal the vital (and vitalizing) conceptual frameworks embedded within all the various forms of indigenous spatial practices and related geographies.

In light of these relationships, chapter 1 discusses several reservation communities where tribal peoples use indigeneity in the material construction of spatial markers. Crafted to parallel the second chapter, which explores White spaces and Indian Villages, this set of research sites demonstrate that Native communities have an equivalent interest in the construction of Indianness via spatial markers, but those markers manifest dissimilar outcomes to White communities. These communities use a variety of strategies in marking tribal space using indigeneity, reflecting the diverse and nuanced senses of identity and history that have shaped each community, even as they respond to similar frames of colonization and racialization.

The second chapter picks up where this introduction leaves off discussing Indian Villages. It documents and analyzes the use of Indianness for crafting White space. Using cartographic and demographic data, I document the twentieth-century proliferation of Indian-themed street names across residential areas in cities, suburbs, and rural towns of every region in the country. We find that while non-White racialized and nonheterosexual space is always constructed as a kind of borderlands delineating the outer boundaries of a "central" normative White space, the spaces that reference Native people dramatically break from this practice and are commonly used where they can directly designate normative White spaces. In contrast to tribal communities and their diverse use of indigeneity, these communities draw from a simplified template without significant variation ("Indian") that operates within the logic of colonialism and multicultural incorporation. These efforts ultimately render Native peoples absent and invisible, and represent a characteristically mundane and concrete example of neocolonial spatial projects.

Chapter 3 turns to two geographically distant sites where the processes of identity and spatial production overlap with one another through a shared reliance on notions of Indianness and, specifically, through a relationship to the historic, real-life Kiowa warrior Set-tainte. The first story centers on the Satanta Day ceremony and the town of Satanta, in rural southwestern Kansas, where a ceremony annually commemorates the town name (derived from Set-tainte) and bestows titles of "chief" and "princess" on successive generations of its residents. The second story considers the Set-tainte descendants' powwow in Oklahoma and broader Kiowa efforts to remember and sanction Set-tainte's anticolonial vision for Kiowa identity and space, as well as continue their traditional maintenance of the Set-tainte name. The comparison and juxtaposition of these stories serves to illustrate the ways Indianness and indigeneity are used in conflicting ways for the production of space, but also explores the possibilities of reconciliation and reconstruction of alternative geographies.

Chapters 4 and 5 engage with Native artists as a way of expanding the scope of Native interrogations with space and its relationship to Indianness and indigeneity. I focus on several artists who have utilized their creative productions to speak to issues of indigenous geography or the constant struggle between the making and unmaking of Native space. I suggest that these works, split between artists that work with the medium of maps (chapter 4) and those that employ public installations (chapter 5), operate through a shared Native relationship to space and colonialism that must privilege the concepts of land and space. I explore how these two mediums offer complementary but also differing modes of centering dispossession, presence, and mobility for Native peoples and communities in a neocolonial nation.

1 Inhabiting Tribal Communities

Clairemont Mesa, California, February 26, 2003

Manitou Way. I find myself steering the 1999 Saturn SC1 that my mother gifted me down this road with an "out of place" name. An Ojibwe word on a San Diego street sign? Longfellow strikes again. I have been noticing it for months. I am supposed to be heading to the local chain store to purchase some household items. I can no longer resist my curiosity. A few minutes off-task now seems insufficient excuse for ignoring this admittedly banal sign. Yet, Manitou Way takes me to Bannock Avenue. Then, Samoset Avenue. Tecumseh Way. Saginaw Avenue. Miami Court. How many streets pull from this Indian theme? I need a map. A few minutes leads to years.

The twisty, curling residential streets and culs-de-sac defy orientation. They flout specificity. The same is true of the street names. Something is hidden here in the mundane. Different names transformed into uniformity. I realize that difference is being made into sameness via green and white rectangles on metal posts.

The local strip mall offers another unexpected guide to the Indian-themed street naming that characterizes this neighborhood. Two stately concrete plaques are mounted on either side of a forgettable exterior corner of the local craft store. Pocahontas Avenue and Rolfe Road come together in a strange act of commemoration. Someone is a historian. Someone is a romantic. I am amused by the prideful matchmaking and historically accurate snubbing of John Smith. I am also suddenly aware of the relationships among culture, geography, power. These signs are narrative. Material. Spatial. These micro-monuments to colonial encounter unintentionally serve as another kind of legend and city map, a residual tool for deciphering this space. They are both—plaque and street sign—markers and mechanisms of the possession and dispossession of lands, of identities.

Grand Ronde, Oregon, March 1, 2013

Saman Uyxət. Salmon Way. I drive through the Tribal Headquarters in Grand Ronde, a relatively new complex I assume to be funded by the economic success of the Spirit Mountain Casino that is situated prominently along the eastern entry to the reservation. On this mild spring morning, the complex is quiet. Few cars travel in or out, although the parking lots indicate tribal employees are stationed in their offices conducting daily business. The complex is small enough to make getting lost impossible, but I miss the parking lot for the education building. I begin executing a Y-turn, carefully, leery of tribal police who might be overeager to turn their attention to strangers violating vehicle codes. As I make the first turn, I find myself facing two lonely signs at a blockaded dead end. Saman Uyxət. Salmon Way. I am still early for my visit with the tribal language specialist, so I decide to explore the complex. After a brief pause and a photograph, I reverse my Y-turn, and continue toward what turns out to be elder housing. More signs, in pairs. Translations of one another. Which came first? What is gained and lost in the pairing? Same names presented differently. I realize that sameness is being made into difference via green and white rectangles on wooden posts.

I am new to Oregon. I do not know much about Chinuk Wawa yet. Or about who precisely was "confederated" on this reservation. Good thing I am preparing to meet with the language specialist. I might get some answers. I find myself again aware of the relationships among culture, geography, power. These signs are narrative. Material. Spatial. This micro-monument to colonial encounter unintentionally serves as another kind of legend and reservation map, a tool for deciphering this space. They are both markers and mechanisms of the possession and dispossession of lands, of identities.

Street Matters

I want my stories about encountering the signs in Clairemont Mesa and Grand Ronde to open a window into the mundane ways that material and spatial productions like street names and signs can operate as and illustrate larger social and cultural tensions. I pair these vignettes together to foreshadow the dialogue between this chapter and the next, which represents the spatial dialogue between different sign systems in different communities.

My first sight of Manitou Way made me wonder what stories informed the creation of these names and signs. How did this name, in particular, arrive in San Diego? At the time, I did not know enough about the influence of nineteenth-century poet Henry Wadsworth Longfellow's *The Song of Hiawatha* (Longfellow 1942 [1855]), which borrows a number of Algonquian and Siouan names and words (some more correct than others) including "Gitche Manito" and "manitos." Dozens of these words, applied to hundreds of sites, have since been put into service as US place-names and within the broader cultural landscape (Vogel 1991). I was far more familiar with the stories circulating around Pocahontas and the academic analysis outlining her ongoing role in settler colonial foundation myths. Her inclusion on these street names and my familiarity with the general mythologizing of Native peoples, however, led me to anticipate how even less-familiar references had more to do with the namers than with those ostensibly being named.

Initially, I looked at the Clairemont Mesa signs as mere markers of arrogance and audacity from a settler colonial society. It smacked of "Indian" mascots being protected and touted with unrivaled passion while Native peoples remain invisibly dispossessed and disempowered. I did not consider that similar signs might also be mobilized in alternate ways or independent forms. I had not seen indigenous-centered place-names being used in parallel ways. By the time I came across the signs in Grand Ronde, I was already collecting examples of street-naming practices in tribal communities.

In the process I have come to appreciate the fluidity and complexity of sign systems, both the physical ones on posts and the cultural structures used to make sense of our world. While I am still critical of the narrative, material, and spatial impositions and settler colonial violences of spatial practices like those found in Clairemont Mesa, I am also equally struck by the power of mundane acts.

I began looking at tribal street-naming to better understand the tensions between varied practices of space-making employing identical technologies. I wanted to know whether these tools that have generally served to help make space abstract could be deployed in the reclamation of submerged, non-abstracted spaces and, specifically, toward the continued practice of indigenous spatialities. How might these physical signs, and the naming practices that created them, indicate the creation of sameness or difference? Henri Lefebvre argues that abstract space is "a product of violence and war" and works to empty out "differences" (Lefebvre 1991, 285). In his seminal work on space, *The Production of*

Space, he argues that capitalism generates abstract space out of the land and out of the intimate relationships people hold with the land. Along with the development of technologies like maps, the abstraction of space produces specific kinds of human-land engagement and, ultimately, enables a universalization of humanity and culture (Craib 2004; Mignolo 1992; Radcliffe 2011). From these frames a series of questions arises: Does Native adoption of this seemingly banal technology reveal the success of assimilation efforts and modes of political incorporation (Bryan 2009)? Can it reflect the persistence of decolonizing spatial imaginings and practices? Does it perform both? Does it vary simply based on scale?

Following these opening examples, this chapter (and the next) is premised on the understanding that geographic markers and marking practices matter. As Lawrence Berg and Jani Vuolteenaho suggest, attention to "critical toponymies" can help us see how "the hegemonic practices of place naming do some of the heavy work of naturalizing and reinforcing the dominance of existing social orders" as well as offer us insights into how "place names represent a contested realm through which people in marginalized societal positions are able to express their own place-bound identities and counter-hegemonic political goals" (Berg and Vuolteenaho 2009, 14).

These first two chapters together offer an opportunity to compare how Native and non-Native communities share, differ, and conflict in their usage of Indianness and indigeneity via street names. These comparisons signpost different spatialities. Such street names and signs are one of the most common and obvious, and yet most thoroughly ignored, makers and markers of space. They are material signals for important spatial constructs and processes. They tell us what is recognized, what is valued, and how. They show us where different aspects of social life can be appropriately and graphically "located" on the landscape. As is the case everywhere they are found, street signs materially stand as labels for the world. They work through the abstractions of culture, identity, and space and then summarize these complexities into a single name, a static graphic, and spatial marker. When street names and signs invoke Indianness or indigeneity, they recall particular sets of spatial relationships. For non-Native communities they serve as evidence of colonization, displacement, settling, and occupation, usually with the intent of rhetorically and textually (but not materially) rejecting such values. For Native communities, they recall homelands and origins, reinforce land-based relationships, intertwine places of creation and knowledge, convey tribe-specific and pan-Indian cultures, and express sovereignty.

The selection of the five sites I discuss in this chapter emerged from research needs that do not lend themselves to readily accessible archives, collections, or research sites. When I began my research on Indian-themed street names in White communities, I was forced to draw on commercial software that allowed me to search for sites by names and that could collect all such instances (Barnd 2010). This approach allowed me to find all occurrences of a street name of interest (say, Cherokee Street). Looking at each of my results, I could then determine whether it appeared as part of a cluster, or not, and assess the scale of the cluster. Finding tribe-specific street names in tribal communities was often even less direct.

The five sites I have chosen (Musqueam, Lame Deer, Grand Ronde, Cherokee, and Dresslerville) offer a diverse geographic range, varied demographic contexts, unique historical needs and interests, and different modes of creating and implementing their signage. They each host a sizeable number of tribe-specific street names reflecting conscious decision-making processes. Together they also provide a good sample of the differences between tribal communities negotiating Indianness and indigeneity as explicit practices in marking space. They were also selected because amid their diversity they share an intentionality that is strongly tied to indigeneity and that thereby suggests the persistence of and continued desire to mark and practice Native spatialities. I do not choose these sites with the intention of exhausting all possible strategies, intentions, or outcomes. Nor did I have interest in creating a comprehensive list of the communities using these tools for marking their spaces. Oftentimes choosing the research sites came about organically, via informal discussions with friends or colleagues aware of community practices (some of which I did not include here). In the end, five cases seemed to encompass most of the variety in signage I found across the United States and Canada without risking redundancy.

Marking Native Space

Producing space is both a conceptual and material process. We encounter our world with sets of values and understandings that shape what we see and how we engage it. In our engagement we then craft our world to reflect our interests. Once we have shaped our environment, we then learn from this world and operate within its constructs. As one can quickly surmise, this is a dialectic process that is dynamic and fluid. It is never complete or fixed, and yet our productions of space generate conceptual and material outcomes that we can view, measure, and contest.

In this light, it is clear that Native space means several things. The term "Indian Country," for example, has both cultural and historic-legal meanings. In the context of the United States, Native space surely reflects a demographic reality. It relates to a land base or a homeland, whether ancient or of more recent design. In the larger sense of space, it describes indigenous meanings given to surroundings and the cultural lenses used in experiencing that relationship between people and the nonhuman world. Native space has a unique set of answers to the questions of what those relationships to the world look like, what meanings are produced, and what identities are generated in the process. As the late Lakota philosopher and scholar Vine Deloria suggests, "Indian metaphysics was the realization that the world, and all its possible experiences, constitutes a social reality, a fabric of life in which everything had the possibility of intimate knowing relationships because, ultimately, everything was related" (Deloria and American Indian Science and Engineering Society 1991, 14). This indigenous frame for cultural and spatial relationships highlights the ways that "power and place produce personality" in a fashion that translates very well to current cultural geography understandings of spatiality.

The use of Indianness found in Clairemont Mesa, California, in the opening narrative (which I discuss in the next chapter) relies on an imagined Other, the Indian—a figure that has a long-standing if fluid function in US and Western discourse. Thus, articulating Indianness through street names and signposts is not equal to referencing Native Peoples. Pocahontas Avenue in Clairemont Mesa is qualitatively different from C'hakćhak Uyxet (Eagle Loop) in Grand Ronde, Oregon, not only in terms of language and content, but also in terms of cultural and community context and the corresponding implications. The difference in choices tells us about each community and tells us about the role Indianness, Whiteness, and indigeneity play in determining who lives there, as well as partially defining the space and those people. The Indianness being referenced and the modes of inhabiting such spaces that are reflected in these two examples are not the same. Both are ongoing processes giving meaning to lands. Both reflect and help generate cultures and identities. One, however, works to produce non-Native space, and the other works to produce Native space and sustain indigeneity.

The figure of the Indian has long proven productive in nation-building (Deloria 1998) and continues to be useful in the overlapping multicultural and "post-racial" eras. In short, the Indian is well-designed for consolidating national identities but also, ironically, for both materially and ideologically crafting White space and "unmaking Native space" (Harris 2002). In non-Native (largely White) communities, the production of Indian-themed spatial markers expresses a colonial ideology and physically marks out the consequences and legacy of anti-Indian spatial practices. The namers use Indianness to mark what are now demographically and discursively White spaces. They evoke the names of Indians to ostensibly reclaim those peoples "vanished" (actually vanquished) from their lands and thereby simultaneously claim those emptied lands. The names, however, represent only ghosts: spectral beings deemed incompatible with, and thus the referential measure for, urban, Western culture. Native peoples are absent except as "insubstantial, disembodied" forms of past resistance (Bergland 2000, 3). When residents of such areas are inhabiting Indianness, then, their signs stand in place of and to some extent even deny tribal survival and indigenous geographies. The residents are not alone, of course. They inhabit in a way parallel to the way the entire nation inhabits Indianness and indigenous space. The residents of Indian-themed places happen to physically reside in locations with referential spatial markers, a phenomenon possible only because of the large-scale and long-term conceptual and material inhabitations, or settlement. But, the entirety of the nation must continually reproduce and naturalize its own construction not only in large ways but also in many small and mundane ways (Billig 1995). In the United States this also requires the ongoing refusal and constant unmaking of Native space.

In contrast, the use of indigeneity found in tribal communities like Grand Ronde represents a collectively imagined Self. In Grand Ronde, names that might elsewhere be seen as obscure or neutral stand within this geography as empowered and potentially empowering assertions of Native presence, tribal sovereignty, and cultural resilience. As Northern Cheyenne scholar Leo Killsback notes regarding place-names, "changing names of the towns on the Northern Cheyenne Indian Reservation, from non-Cheyenne names to names rooted in Cheyenne culture, is an assertion of nationhood" (Killsback 2005). Likewise, spatial markers offer material examples of the co-constituency of identity, culture, and space. These humble bits of text and steel highlight the relationship of names to community identities, and in the case of tribal communities, they frame and recall the tensions between colonization and sovereignty, history and culture, representation and race. For Native communities, such names and signs can effectively represent and function as one of the many mundane but necessary practices of sovereignty.

Rather than re-narrating and building on the legacy of dispossession, the names and signs in these communities serve and seek to extend control over tribal life, and to reshape or reclaim geographies away from a violently non-Native worldview. They testify to tribal survivance (Vizenor 1994). Articulations of indigeneity via tribal community street names help simultaneously create and reflect ongoing constructions of Native space, sometimes even including spaces that may not be under official tribal control or that might extend beyond the broad, multiple, and overlapping realms that make up Indian Country (Biolsi 2005).

I try not to make the enticingly simple claim that demographics and politics drive street-naming practices, but I will say that tribal decisions on these matters must necessarily reflect the complicated interactions between culture, politics, and context. These factors matter to tribal communities precisely because space is so fraught with tension, and tribal space in particular is necessarily implicated in struggles over land use, land claims, and relationship to homelands. In lands that are by definition continually subjected to colonial interventions and restrictions, these signs announce mundane practices of sovereignty rooted in inherent self-determination. They represent one small act within and toward a comprehensive practice of selfhood. Of course, the intentions and usages are not aligned across these communities. What all have in common, however, is their functionality as bold markers of ongoing or reasserted Native geographies. Taken together, these examples from different Native communities reflect a spectrum of approaches, intentions, and tactics in the practice of Native space.

Signs and Designs

The decision to employ street names and signs is not necessarily a given. I briefly outline some of the factors influencing these communities but spend most of my attention on the spatial implications of the final products. This research choice reflects, partially, the challenges of tracking down decisions made, in some cases, decades ago with little fanfare and less documentation, as well as the fact that space-making is a contested, fluid, and ongoing process without static purpose or understandings. Thus, my approach is only partially interested in explaining how or when tribal street name choices were made, and more interested in how they matter now that they exist. More humbly, I also simply hope to draw attention to this unique spatial aspect of indigeneity and its practice via one specific but increasingly common material expression.

Hovering over the importance of materiality for a moment, I want to briefly underscore here that my interest lies with the physical street signs just as much as with the names themselves. While the signs could be oversimplified as purely the material manifestation of names, the physical implementations and installations of signage actually provide for a great deal of additional analysis. Tribal communities have numerous options when deciding to install street signs, each of which reflects, shifts, or generates additional meanings and kinds of spaces. On the practical side, communities must answer a number of questions: for example, How will streets be named or renamed? What purposes do the names or the signs serve? How do they wish to design and present the names on the physical signs (if signs are posted)? Will the names and/or signs be monolingual or bilingual? If bilingual, which names will be primary and which supplemental? Can such distinctions be mediated differently, or more equally? Does the community want bilingual names to appear on the same sign or on separate signs? In either case, which name will be placed above and which below (if stacked)? Which orthography system(s) shall be chosen? Table 1.1 below offers a quick accounting of the different choices, within just the five communities discussed in this chapter. While the differences may at first glance seem trivial, I argue that the variations lead to a number of outcomes and implications and, more importantly, that the choices clearly matter enough for increasing numbers of indigenous communities to take up and implement these decisions across a number of contexts.

Tribal self-representations via street names and signs is not practiced everywhere, but does seem to be growing as communities develop their residential and government complexes and diversify handling of cultural, economic, and other infrastructural aspects of tribal life. Tribal headquarters and other centralized resource structures (health care centers, courts) are common starting points for implementing indigenous-specific names and signage. These are often new construction sites that present prime opportunities for asserting tribal presence, sovereign authority, and cultural identity. Talequah, the Cherokee headquarters in Oklahoma, for example, posts bilingual signs (in the Cherokee syllabary) running six blocks along Muskogee Avenue between Chickasaw Street and Morgan Street. Window Rock, Arizona, the capital for the Navajo Nation, features a

TABLE 1.1 Street name and sign choices in five tribal communities

Site	Street name language	Suffix Language (Drive, Street)	Sign type	Name Placement and Relative Size	Orthography	Example
Lame Deer, MT	English	English	Single	All centered	Latin only	Yelloweyes Drive
Musqueam, BC	hǝṅ̓q̓ǝmiṅ̓ǝṁ & North American Phonetic Alphabet (NAPA)	English	Single, shared space	hǝṅ̓q̓ǝmiṅ̓ǝṁ below, NAPA (superscript) above	Latin & NAPA	Yuculta Crescent/ (NAPA)
Grand Ronde, OR	English & Chinuk Wawa	English & Chinuk Wawa	Dual, separate space	English above, Chinuk Wawa (subscript size) below	Latin & IPA	Dragonfly Drive/ Ulq-later-inǝpʰu Uyxet
Cherokee, NC	English & Cherokee	English & Cherokee	Single, shared space	English above, Tsalagi below	Latin & Cherokee syllabary	Tsali Boulevard/ ᎬᏆ ᎤᎬᎭᎴᎠᎤ
Dresslerville, NV	Washiw & English	English	Single, shared space	Washiw above, English below (in parentheses)	Latin only	Patdul'Negeeh Way / (Eagle)

common mixture of English street names like Main Street, Tribal Hill Road, Townhouse Circle, and Mustang Road posted along with tribe-specific names like Toquema Boulevard, Kaibeto Drive, and Chee Dodge Drive.

Choosing Tongues

In Lame Deer, Montana, the street names and signs are entirely in English. This could easily be read as the least tribally assertive of my examples, if we rely only on linguistic presentation as the most important indicator. Yet Lame Deer is also one of the more demographically, geographically, and culturally protected Native communities you can find. The community is 95 percent Native (overwhelmingly Northern Cheyenne), and relatively culturally distanced from most of the non-Native communities nearest them. Despite the English-language signs, the names solidly reflect Cheyenne history and identity. In British Columbia, on the other hand, the Musqueam Band (Reserve 2) provides only Anglicized spellings of their language. This approach suggests their signage might be the most motivated by culture and politics, and offers perhaps the clearest example of tribal and spatial activism (at least via street signs). The Musqueam community is undeniably affected by its urban surroundings and its history of a multiethnic resident-lessee population, but it also has a notable tradition of protecting sovereignty and Native space. The communities of Grand Ronde (Oregon), Cherokee (North Carolina), and Dresslerville (Nevada) all employ both English and tribal language or writing systems. Each of these bilingual sign systems, or what Tim Edensor calls a "signscape," illustrates decisions to balance everyday practicality and cultural-political intentionality (Edensor 2004). Together, these five sites indicate a (sometimes only implicit) tribal recognition that indigeneity has a spatial dimension and that tribal spatiality can be expressed in multiple, but still effective and concrete, ways. Despite or perhaps precisely because of the very banality of tribal street sign systems, these material expressions also seem to effectively reinforce the scope and nuance of what it means to practice tribal sovereignty and produce Native geographies.

Comparing the five sites, language choice and presentation range widely. As alluded to above, the sign systems in Lame Deer and Musqueam draw primarily from a single language source (English and hənq̓əminəm̓, respectively) even as they "code switch" through cultural references or language. Lame Deer includes streets named after Chief Dull Knife, for example, while Musqueam includes English language qualifiers on street signs (i.e., Avenue and Drive). The main street in Lame Deer, for example, is Cheyenne Avenue. Following this model, more than twenty street names and signs in this town and several of those found throughout the larger Northern Cheyenne reservation landscape are relatively modest in marking it as an indigenous geography. Unlike the street names and signs found in some tribal communities, those found in Lame Deer have not been reworked to emphasize tribal language(s). As a small, relatively homogeneous community in southern Montana it predictably both suffers and benefits from an informal cultural, economic, and geographic isolation.

Lame Deer serves as tribal headquarters for the Northern Cheyenne people. Home to a few thousand residents, this rural town is located within the larger 444,000-acre reservation. The Tongue River Reservation was established in 1884, expanded and renamed the Northern Cheyenne Indian Reservation in 1900. The Cheyenne settled here after most of the tribe had been forced across the northern Plains and into Indian Territory. In 1878, Chiefs Morning Star (Dull Knife) and Little Wolf fled Indian Territory with nearly three hundred people to begin a 1,500-mile journey back to the northern plains. The pain and simplicity of this "hazardous" return through homesteaders, ranchers, railroads, and troops is beautifully expressed in the tribal biography produced by Chief Dull Knife College. The editors explain in understated fashion that when the Northern Cheyenne left Indian Territory, "Tsehne'evahoohtoosemevóse Tsetséhestáhese, they were going home" (Ambler et al. 2008, 26).

Although this community is insulated, its signage is relatively accessible to English readers in terms of being composed of words that are familiar and readily pronounceable, even if the deeper cultural references and significance are not clear or fully available to most non-Cheyenne. This allows Native and non-Native alike to access these signs, since comprehensive tribal language use is still under (re)development. As figure 1.1 suggests, many of the names found in Lame Deer could theoretically be found in any number of places, especially examples such as Bixby or White River. What makes them Northern Cheyenne–specific is not the linguistic representation. Rather, the names reference cultural and historic touchstones not found outside of this community. Most of the street names in Lame Deer and some around the reservation reflect their placement within a Northern Cheyenne geography and community: Sweet Medicine Road, Medicine Walking Woman Circle, Stands-in-Timber Drive, Medicine Elk Drive, Yelloweyes Drive, Dull Knife Drive. Similar names and signs likewise mark local subdivisions: Crazy Head Heights, Two Moons Heights, Eaglefeathers, Big Back, Shoulder Blade Heights (Foote 2010). These historically derived name choices fit into an intentional integration of tribe-specific histories and titles with the landscape, and match broader efforts at tribal cultural protection and promotion.

Although these street names have functioned for many decades, there are indications that the names might change in the future. In 2005, then tribal president Eugene Little Coyote articulated a desire to change the town names of Lame Deer (to Black Lodge) and nearby Busby (to White River) in line with their traditional and commonly used names (Bohrer 2005). In 2007, the Tribal Council's Cultural Commission intended to put forth a resolution calling for greater implementation of a 1997 ordinance. That ordinance asserted the Northern Cheyenne language as the official language of the Northern Cheyenne people. The commission wanted the Tribal Council to enact two changes to increase efforts to promote the Cheyenne language: to produce bilingual resolutions and ordinances, and to "convert community and street signs from the English language into the Northern Cheyenne language with accurate translation, therefore prioritizing the Northern Cheyenne language and promoting the language to the Northern Cheyenne people" (Northern Cheyenne Cultural Commission 2007). The commission's formal presentation was apparently preempted when the tribal council president was ousted from his position, but that group clearly saw importance in further expressing a Cheyenne-specific Indianness through the official markers of tribal space. The resolution has not since been presented or adopted.

One counterintuitive reason for non-adoption of this intended proposal could be a desire to maintain tribal cultural and spatial integrity. Many of the current street names were selected decades ago, although signage has been minimal or sporadic and little used in daily life. Until 2010, the street names in Lame Deer were so insular to this tribal community that many of the roadways simply had no signs (sometimes no names). Many still lack signs, especially the unpaved residential roads. Tribal members already knew their community landmarks and used informal navigational means. A 1964 planning document noted that besides the "oiled" Cheyenne Avenue "the other streets were little more than trails," and thus proposed establishing new streets and improving existing streets and subdivisions (United States Bureau of Indian Affairs 1964, 9). A 1969 transportation study noted that Lame Deer had "no officially registered plats," which suggests that roads were, like most of the homes, still improvised as needed, rarely recorded or submitted for approvals, and used without names or signs (Meyer 1969, 1).

FIGURE 1.1 Sample sheet for street signs in Lame Deer, Montana. Courtesy Newman Signs.

In 2010, the Northern Cheyenne secured a federal infrastructure grant to upgrade transportation to facilitate the emergency response system, ease delivery of services (such as parcels and utilities), and simplify directions for visitors. Using humor to drive home the point about navigation challenges for outsiders (the late) Special Projects Coordinator A. Clifford Foote explains that

> a lot of times you hear someone say "you go down two blocks, turn left and it's the third house on the right. It should be a green house with a black and white dog sitting at the corner." Well over the summer the house got painted and it's now blue and the black and white dog has w[a]ndered off and now lives across town and you are not sure if you are at the correct house. (Foote 2010)

These are common concerns in many rural reservation communities. Rod Johnson, manager of the K'ima:w Medical Center's ambulance department serving the Hoopa and Yurok tribes in northern California, indicates their rescue crews were often dependent on the level of familiarity of local dispatchers.

Paralleling Foote's story, Johnson explains that dispatchers were sometimes narrating a turn-by-turn local geography: "Go three roads down from this intersection and then take a right at the fork in the road by the old truck" (Korns 2012). Johnson likewise notes the ambiguity of such fluid and dependent navigation techniques: "One time [the dispatchers] told us 'you'll see the chickens in the yard, turn right there'" (Korns 2012).

Johnson and Foote clearly draw on humor for a challenge to efficient service delivery, yet Foote also uses a more serious approach, going so far as to equate the creation and protection of street signs with an overall sense of tribal pride and community duty. Noting that previously installed signs were now missing, he pleads, "Let's not forget what sacrifices Little Wolf and Dull Knife made so that we can have this reservation we live on" (Foote 2010). Invoking the names of the two most revered tribal figures, both of whom led the Northern Cheyenne back to Montana from Oklahoma to reclaim homelands, Foote seeks to characterize those not actively supporting the infrastructure project as disgracing the undeniable courage and sacrifices of their elders and as fundamentally disrespecting their hard-won homelands.

Against both Foote's sentiment and my own contention that tribal community street naming and signage offer unique examples of the production of space—specifically, Native space—I want to suggest that the missing signs in Lame Deer can also be read as acts of space-making, of another sort of indigenous geography. The federal infrastructure project leaders indicated that the creation of the street names (where needed) and signs (often needed) would ease the process of securing driver's licenses, setting up bank accounts, purchasing vehicles or rifles, and completing other exchanges that demand designation of a physical address (Foote 2010). For all of the espoused benefits of having street names and signs, many tribal communities may also read such markers as tools for control, or tracking. While they can certainly help deliver services to those who want them or facilitate tribal empowerment in many ways, they can also make the community and its individual members subject to the same kinds of mapping practices that surveyed their lands before and during occupation (Bryan 2009; Craib 2004; Rafael 2000; Stark 2012; Warhus 1998). Mapping activities parcel out discrete segments for private property, establish reservation boundaries, and generally make tribal landscapes "legible" to others. Street names and signs, and the general effort to mark land, can render Native peoples and lands locatable, or mirror the same use of spatial rationalities effectively deployed to manage large, diverse urban populations in cities (Rose-Redwood 2009). As Lefebvre's under-standing of abstract space suggests, such signs can materially signal the efforts to abstract land and subject people living on that land to a parallel abstraction that works against traditional Native spatialities. Thus, we must acknowledge that naming and signage, like all technologies, can be used for competing interests and in oppositional ways.

As history shows, however, a refusal to participate in the dominant technol-ogies does not guarantee escape from them. This does not suggest acquiescence is the only option, but it does highlight one of the greatest lessons learned from conflicts between settler colonists and Native peoples. In the context of unequal power dynamics, creative adaptation or co-optation of imposed practices even-tually tend to prove far more effective means of resistance than outright rejec-tion or opposition. I would argue that this is why so many tribal communities have chosen to take control of their street names and signs and participate in mapping rather than eschew them completely. Similarly, few Native people still see an irreconcilable contradiction in using non-Native technologies for Native purposes (if those categories can even be so easily separated and maintained). Despite the dangers of incorporation and appropriation, establishing clear land boundaries and jurisdictions can certainly also serve to help establish a degree of autonomy, facilitate economic development, define tribal identities and activities, and ultimately even provide a base for further land reclamation (Bryan 2009).

Using these impositions as tools and opportunities, it seems clear that tribal communities have made every effort to craft laws, policies, and practices in ways that reflect and, ideally, benefit their local cultures, identities, and histories. Certainly some Native communities or individuals have simply mirrored the values and practices of the dominant non-Native society or even used them for undue personal or family gain. Most have found creative means to balance or match the benefits of technology and policy with the complex and ever-changing cultural and political needs of their communities in ways that respect and pro-tect fluid tribal interests.

Given the relationship among culture, identity, and space, it would not be sur-prising to see the Northern Cheyenne Tribal Council Cultural Commission's proposal resurface and be adopted in the near future. On December 15, 2008, for example, the Tribal Council passed a resolution (which became Montana House Bill 412) proposing to change the names of Squaw Creek and Little Squ*w

Creek—which I dis/spell here following C. Richard King's example (King 2003) (east of Lame Deer) to North Woman and Iron Teeth Woman Creek. Clearly the Northern Cheyenne leadership understands that naming gives meaning to both the namers and the named—not a trivial matter, nor beyond need of their attention. Lame Deer's street names and signage (if they survive!) might soon reflect parallel changes elsewhere that seek to fully integrate cultural and linguistic practice into everyday life. We continue to hear from leaders in tribal cultural revitalization that linguistic development is one of the key tools needed to make their efforts sustainable. The late Blackfeet linguist Darrell Kipp, one of the most venerable figures in tribal language reclamation efforts, was emphatic in arguing the importance of language for a holistic cultural recovery. "The quintessence of a tribe," he frequently asserted, "is found in their language, not their culture" (Kipp 2007, 106). Kipp's stance may not necessarily preclude substantial cultural or spatial recovery and reinvention without the thorough use of traditional languages, but one certainly cannot dismiss the great value it does bring when implemented.

Nation Under Par

In contrast to Lame Deer, the Musqueam Nation (Reserve 2) is an urban reserve. Its location alongside (ostensibly "within") the city of Vancouver seems to require diligent protections, including tribally and linguistically explicit spatial markers. The Musqueam crafted their sign system entirely in the Musqueam (hən̓q̓əmin̓əm̓) language. The physical street signs posted since the 1960s use both a Latin alphabet and the North American Phonetic Alphabet (NAPA). The Latinized words are prioritized on more than twenty individually named streets, with many including a supplemental phonetic orthography (NAPA) placed in a sort of superscript (see fig. 1.2). The names and signs used on this reserve are unique illustrations of a cultural and linguistic approach to expressing indigeneity and crafting an indigenous geography. Both the Latinized and NAPA versions are transliterations, and not translations; thus the signs offer no English equivalent of the Musqueam words (besides the Avenue and Drive designations). Those who can read NAPA are able to approximate the hən̓q̓əmin̓əm̓ word sounds. This still leaves most of the streets "illegible" to non-Musqueam speakers, except, one would hope, Salish Drive and Musqueam Avenue. Beyond those two exceptions, non-Musqueam passersby or map readers confront names and signs such as Thellaiwhaltun Avenue, Ke Kait Place, Staulo Crescent, and Kallahun Drive with little access to the spatiality being signalled.

FIGURE I.2
Musqueam (Reserve 2)
street sign. Photo courtesy
Harley Augustino.

In the United States, where reservations rarely intersect with urban spaces, English signs or translations accompany most if not all tribal language sign systems. Considering Lame Deer and Musqueam together, then, raises the question of whether proximity to, and therefore the cultural threat of, urban centers might increase the likelihood of tribal communities choosing indigenous language for public activities and presentations. This seems especially likely given that near-urban groups are heavily affected spatially by external impositions and non-Native expectations of recognizable "Indian" authenticity and identity. The Musqueam Reserve 2 lands have a long history of being designated for economic development, and thus leased on fixed-terms to non-Natives and subsequently used for golf courses (violating treaty rights), staging city housing developments, and supporting Chinese farmers seeking opportunities denied in the propinquous city. According to the 2006 Canadian Census, Musqueam Indian Reserve 2 was home to 1,371 residents, including 605 of whom (or 44 percent) identified as Native (most being Musqueam).

In the 1960s, Musqueam chose increased economic revenues allowing housing developments, and leasing homes to the general population. Yet, they did not concede their street names. As local historians Tom Snyders and Jennifer O'Rourke point out, when city officials proposed naming the Musqueam Park development streets near the Shaughnessy golf course after famous golfers, the band leadership refused and insisted on hən̓q̓əmin̓əm̓ names reflecting Musqueam culture and geographies. They rejected suggestions of Demaret,

Colk, Casper, and Locke in favor of exclusively using their local Central Salish language for tribally relevant names instead: Halss, Kullahun, Semana, and Tamath, respectively, for the streets in question (Snyders and O'Rourke 2001, 117). Beyond the use of language, a subset of these choices consciously represent "ancient settlement[s]" on or near the current reserve site. This seemingly simple naming act undeniably insists on recognizing, marking, and practicing Musqueam geographies.

Within the context of a Native space persisting alongside and underneath a colonial city, Musqueam has actively decided to engage in activities that are deemed beneficial for the reserve and its people. Like all tribes, they seek to utilize diverse methods of survival in a larger capitalist society to (in part) subsidize other cultural and political projects, including language-reclamation programs, fishing rights, and land claims. Their cohosting of the 2010 Winter Olympics and a published *Community Profile* both attest to the fact that Musqueam believes "revenue is critical to becoming a self-sufficient Nation" and that "with revenue we can greatly improve the life of all members" (Musqueam Indian Band 2007, 33).

Yet, this particular band has been at the center of a number of legal challenges, including the Supreme Court of Canada's 1984 *Guerin* and 1990 *Sparrow* decisions. *Guerin* confirmed Canada's federal fiduciary responsibility to First Nations, precontact rights were legally affirmed, and the Musqueam became the first Native nation to successfully sue the Canadian government. In the *Sparrow* case, the court recognized tribal member Ronald Sparrow's ongoing fishing rights via First Nations' treaties. These rulings led directly to larger political and legal affirmations of tribal sovereignty in Canada, including protected resource rights. The Musqueam are transparently keen on taking advantage of economic opportunities while simultaneously pressing issues that challenge Canadian cultural and political hegemony and support tribal interests. As the Musqueam street sign system and these court cases illustrate, this community has engaged in multiple modes of protecting, sustaining, and projecting tribal geographies.

Subtitled Spaces

So far, I have shared examples from Lame Deer, Montana, and Musqueam, British Columbia, sites that offer very different linguistic approaches to self-representing via street names and signs. In both cases, and in the cases that follow, the comprehensiveness of the sign systems take on equal importance with the content and linguistic choices. They also use a tribe-specific and place-based identity (fluid and locally specific as it may be) to claim both particular lands and a more conceptual and extra-reservation sense of their respective Native geographies. The simple act of exceeding official boundaries disrupts what Mishuana Goeman calls a "settler grammar" or the mundane "systems of rules [and] indexes" that "give settler place meaning and structure" (Goeman 2014, 237).

Reservations are supposedly what remains for tribes "outside" of the settled/occupied/colonized American landscape. Yet, street names and signs materially resist the colonial grammar that still textually marks many tribal landscapes, either by replacing or operating in coexistence with tribal grammars of place. As an explicitly articulated Native space inevitably defies, shifts, and grows beyond its "official" bounds, it unsettles colonialism and conquest. Such unofficial and unsettled space is thereby open to reinscription and new spatial imbrications precisely where it was supposedly fixed and resolved—spatial projects that treaties and then plenary power ostensibly accomplished or continue to accomplish. While reservation street names and signs themselves remain within established colonial bounds, the act of generating names and signs and otherwise consolidating culture with concrete markers nevertheless works with and within efforts to, for example, reclaim resource (co)management responsibilities, off-reservation lands, and sacred sites, and to craft new economic ventures, as well as to further the movement and growth of peoples and cultures. As reflections of their cultural context, the street names exceed the containment of the reservation. They mark the ways that Native space spills out past reservation lines. They simply provide a graphic representation of the indigenous geographies already in practice.

With this in mind, for the next three examples the physical signage becomes even more obviously valuable as an object of study toward understanding community spatial self-imaginings. In terms of the placement and presentation of the names and signs, those examples found in the communities using bilingual sign systems are perhaps the most illuminating. The Confederated Tribes Grand Ronde, Eastern Band Cherokee, and the Dresslerville Colony each offer examples of self-conscious and assertive spatial productions in which the choice of languages used on street signs continues to matter, but so does the placement of the signs and text in relation to one another. My analytical frame therefore adds an attention to form or design, as well as to content and language.

These three Native communities all use bilingual signs but also produce a number of different sign systems. Located in western Oregon, the Grand Ronde reservation is now increasingly visited by those travelling to the tribal casino, Spirit Mountain. Grand Ronde utilizes the International Phonetic Alphabet (IPA) orthography to transliterate the Chinuk Wawa language. Unlike Musqueam, however, most of the Chinuk Wawa street signs include English translations. The handful of street signs that cover the elder housing, tribal housing, and administrative complex all provide both all-English and all–Chinuk Wawa names, such as Blue Jay Court/Qeysqeys Uyxət and Dragonfly Drive/Ulq-latet-inəpʰu Uyxət. In these instances, Grand Ronde chose to supply two names on separate signs, which suggests a parallel existence for the languages (see fig. 1.3). The only two outlier streets are in the Family Housing complex; Tilixam Circle (people, or family) and Tyee Road (chief, or important person).

In 1954, the US Congress officially terminated Grand Ronde, deeming them sufficiently assimilated and racially amalgamated to become individual citizens. After more than a generation of activism and lobbying, however, the tribe gained restoration in 1983, and four years later officially reclaimed tribal stewardship over 9,800 acres of land. Following Daniel Wildcat and Raymond Pierroti's argument that "being indigenous" requires and reflects a long period of time and a developed relationship to a specific land, the Chinuk Wawa street names and signs can be read as an important Native spatial act for a relocated and confederated tribe needing to slowly solidify its own presence, to geographically (re) locate and to establish indigeneity (Pierotti and Wildcat 2002).

It is worth noting that outside of the casino patron circulation, tourism is not a significant element of the Grand Ronde reservation. Like Cherokee, North Carolina (as I will show), the overlapping and contested sets of writing systems, languages, cultural contexts, historical memories, and political tensions impact the Grand Ronde geography. Longtime Grand Ronde linguistic consultant Henry Zenk suggests that confederation forced the tribes toward the promotion and development of Chinuk Wawa, an "intertribal hybrid language" that developed in the 1800s out of the more well-known and regionally shared Chinook trade jargon (pidgin) once used across what is now the Northwest (Chinuk Wawa Dictionary Project 2012, 13). Like many of the peoples in the far west of the continent, Grand Ronde exists as a conflation of culturally and linguistically unique peoples confederated as one "tribe." This reservation became home to twenty-seven bands and tribes relocated from throughout western Oregon, southern Washington, and northern California. Without a single prominent traditional language, Chinuk Wawa served both as "general communication within the founding reservation community" and as the best way of "organizing themselves as a wholly new community, and undertaking a wholly new way of life" (Chinuk Wawa Dictionary Project 2012, 16). In the introduction to the latest Chinuk Wawa dictionary, the editors point out that recorded elder speakers clung to Chinuk and passed it on through family circles. Speaking the language had soon "taken on a symbolic significance" that enabled its continuation and now revitalization (Chinuk Wawa Dictionary Project 2012).

While some of the confederated peoples had a more sustained, traditional engagement with the assigned reservation lands, others needed to deepen or initiate new relationships. In the newly formed and shared reservation space, all relationships were affected and transformed, and in turn the peoples were changed too. In all, the successive experiences of displacement and termination, as well as the confederated nature of Grand Ronde, create a context in which street names and signs no doubt function as additional tools for cultural and linguistic revitalization and continuation, as well as for more fully claiming and establishing Native space not originally (except for those bands from that valley) of their own making.

FIGURE 1.3 Chinuk Wawa and English street signs at Grand Ronde, Oregon. Photo by author.

As this set of community signs helps illustrate, the simple graphic presentation of a road sign is embedded with spatial processes and implications. Each example provides a glimpse into the unique sets of circumstances and strategies in a given tribal community. By virtue of their public function and purpose, bilingual sign systems inherently and publicly help tribal communities negotiate the relationships among culture, language, and geography. They textually and materially place writing systems, languages, cultural contexts, historical memories, and political tensions in conversation. They officially signal multiplicity and simultaneity, and because they are geographic markers they also announce contending and overlapping spatial constructs (Massey 2005).

Overlapping geographies necessarily produce inclusion and exclusion. The Cherokee, North Carolina, use of the Cherokee syllabary offers an interesting illustration of the tensions of spatial inclusion and "exclusion." In the town that serves as tribal headquarters for the Eastern Band Cherokee, the street names and signs use a complex combination of English, Cherokee, and the Cherokee syllabary. Many of the town street signs, for example, are made up of English names with Cherokee syllabary translations, such as Rhododendron Trail/ꌚ ꭱꭲ. Others, like Tsali Boulevard/Ꮳꮅ ꮫꭺꮧꮒꮻꮪ, feature a Latinized script for a Cherokee-origin name (in this case, commemorating an important Cherokee person, Tsali), an English identifier (Boulevard), as well as the Cherokee syllabary translation. The Eastern Band script used so widely in this tourist destination is the only one in the United States that offers a local, tribe-specific orthography, although we find Cree syllabic parallels, for example, in Canada.

As with the Musqueam utilization of hən̓q̓əmin̓əm̓, the graphic presentation of the Cherokee syllabary on street signs prioritizes Native language speakers and language maintenance practices, rather than "outside" sign readers. Legibility of the syllabary is limited both within and outside of the Eastern Cherokee community, even as the Eastern Band is invested in a number of efforts aimed at regaining Cherokee language fluency and literacy. The Cherokee sign system, however, eases exclusion of "outsiders" and nonreaders by posting both English and Cherokee as translations of one another. The syllabary signs were added relatively recently (in the mid-1990s) and indicate the decision to supplement rather than replace names and signs. In this way these practices mirror those of Dresslerville and Grand Ronde and separate this system from Musqueam's firm and more internally focused sign system. The eased literacy for the majority of travellers in the town of Cherokee is also highlighted by the placement of the English text, which appears in the standard Helvetica font and at the same position as signs

FIGURE 1.4 Street sign in
Cherokee, North Carolina.
Photo courtesy Amanda Green.

found in streets across the United States. In fact, despite its potential for exclusion, the syllabary actually appears as more of a subscript, as it is placed below and slightly smaller than its English counterpart, where it can be read as supplemental or additive rather than central. Again, the Musqueam model provides the most dramatic counterexample to this design choice.

Given the importance of tourism in this community, it makes sense that inclusion of the syllabary at least partially targets visitors seeking a "cultural" experience. Indeed, anthropologist Margaret Bender notes that the largely "illegible" (to outsiders) syllabary adds to a sense of authenticity experienced when visiting sign readers encounter indecipherable texts. "Writing is most exotic" she tells us, "when it just barely works for the viewer as an icon of meaningful writing" (Bender 2002, 143). In other words, viewers will understand the syllabary as an orthography, but since it escapes their comprehension, the script takes on greater symbolism as a marker of difference. For the purpose of this analysis, an overlooked aspect of that perception and construction of difference is that it extends to the space itself, as the graphics help to simultaneously reveal and imbue the landscape with an "authentic" Indianness and thus an inaccessible set of meanings. The syllabary effectively works to estrange non-Cherokee (and perhaps some Cherokee) from the geography and to simultaneously embed the Cherokee and the land within one another. In many ways this proves true whether we approach the text from a Cherokee or non-Cherokee positionality.

The fact of tourism and the Eastern Band's strategic investment in tourism since the 1940s does not exclude the crucial role cultural and linguistic revitalization play in the presentation of the syllabary. The use and importance of Sequoyah's famous script obviously extends long before street signs and tourist traps. This highlights what is often an ironic and unsatisfying, but sometimes symbiotic and productive, relationship between tourism and indigenous culture (Bryan 2009). Well before its mid-1990s implementation on street signs, the syllabary served as an important cultural symbol within the community. In her extensive work on the syllabary in Eastern Band Cherokee and particularly with those in the town of Cherokee, Bender notes that use of the orthography affects all facets of tribal life. Stretching back generations, she tells us, "the syllabary was clearly a key component of the semiotic landscape of the [Qualla] Boundary and its environs. It played a key role in education, the media, tourism, religion, political activism—and in every important dimension of Eastern Cherokee life" (Bender 2002, 34–35).

In contrast to the model used in Cherokee, other communities use bilingual signage that raises different questions about the tensions and overlapping of geographies. The Dresslerville community in rural Nevada is a Washiw reservation established in 1916, long serving as tribal cultural epicenter. It hosts just over a dozen street names for its 250 residents, using combined English- and Washiw-language names on a single sign, with names like Memdewee Run (Deer) or It'Mahowah Circle (Cedar).

If we rely on a straightforward translation of the street names, we find some of the "same" names used in non-Native communities. Dresslerville hosts several streets named for "nature": Deer, Cedar, Eagle, Juniper. These are shared street names that one might note are not even particularly unique to these towns, or to this region. A non-Washiw speaker might encounter these signs and not be impressed or even unsettled by their unfamiliarity. These references, however, point to significant cultural context and meaning differences held within tribal stories, songs, and cultural uses of deer, eagles, cedar, and juniper. And we might still return to reconsider the necessarily complex and entrenched self-generated meanings of Wa She Shu for the Washiw themselves (see fig. 1.6).

As these examples suggest, the Dresslerville sign system uses only the Latin alphabet. Washiw designates the street name, English the suffix. An English translation of the name follows. This bilingual approach falls somewhere in between the approaches used in Lame Deer and Grand Ronde, using a Latinized script, and English, and providing direct translations. The Washiw approach, however, also takes a strong linguistic stance that is sustained even as the Washiw names are simultaneously doubly bilingual. In other words, Wa She Shu Way (The Washoe People) obviously includes the qualifier "Way" in English rather

FIGURE 1.5   Intersection of Pba'ul Street and Patdul'Negeeh Way street signs in
Dresslerville, Nevada. Photo courtesy Helen Fillmore.

than a Washiw parallel. Yet, the Washiw language gains strategic prominence
by means of Dresslerville's placement and presentation of English and Washiw,
and how these seemingly neutral or trivial choices have additional implications for mapping (which I address shortly). Although the "translation" of the
Dresslerville street names generate a seemingly common set of words, the act of
translation here reminds us that this naming/signage firmly rests on a specifically
Washiw naming system and cultural foundation.

In Cherokee, North Carolina, both the English and tribal (Cherokee) names
are displayed on the same physical sign, just as in Dresslerville. Yet it seems
the manner in which a tribal street name is positioned and scripted on a sign
changes how non-Natives adopt or erase them, particularly in the context of
digital cartography and official maps. As noted, the Cherokee syllabary translation is located beneath the English name. The placement above and below does
not inherently convey significance, even if Western standards commonly rely
on positional understandings of hierarchy where above (the top) is often more
valued and powerful than below (the bottom). Yet, in the move from graphic/
physical sign to digital map, some tribal communities lose cartographic representation of their name choices.

FIGURE I.6 Wa She Shu Way street sign in Dresslerville,
Nevada. Photo courtesy Helen Fillmore.

The Dresslerville naming system, which features a Latinized orthography and
centralizes the Washiw language, results in their street names showing up on
Google Maps, as well as US Census maps and other official documents (see fig.
I.7). Searching online digital maps easily produces results for Wat'Shemu Way,
It'Mahowah Circle, Pba'ul Street, and Boddu'h Way. Searching for the English
translations of these Dresslerville street names (Main/Carson River, Cedar,
Dark Brush, and Elderberry), however, usually produces no results. Musqueam
generates the same outcomes, since they offer mapmakers no choice but to use
the names they present or to subject readers to the North American Phonetic
Alphabet. Google maps do not, as we might predict, include any of the Chinuk
Wawa names for the Grand Ronde streets since they are rendered only in the
International Phonetic Alphabet supplemental text. The search engine's digital
maps likewise never feature the Cherokee syllabary. Searching for those street
names, even if using tribal orthographies, yields no results.

There is probably no coherent guideline or set of practices for how mapmakers determine what to prioritize when confronted by these sign systems, but it is more than likely that a combination of name placement on the sign and orthography lead non-Native mapmakers to read many tribal language names as supplemental and thus not include them in their renderings. Scale matters here as well, since numerous world languages and non-Latin writing systems are produced for international maps, including Hangul (Korean), Kanji (Japanese), Arabic, Nepalese, and Cyrillic (Russian). One could reasonably argue that the language choice, orthography, and placement of the Chinuk Wawa and Cherokee names underneath the English confirms not just the greater practicality of English and the "secondary" nature conferred to tribal languages, but also its dismissal by cartographers as too localized to merit renderings and verifications.

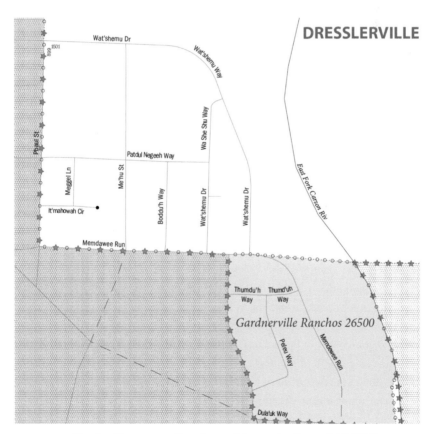

FIGURE 1.7   Dresslerville map. 2013 Boundary and Annexation Survey BAS, US Census.

In the case of Dresslerville, the Washiw language-implementation strategy submerges English. Contrasted with other tribal communities across the United States that also have bilingual signage, the Dresslerville street signs visibly prioritize Washiw words by adding the English translations underneath, and within parentheses. The use of parentheses unmistakably signals that the word included within is secondary or an "also known as" reference, and thus exists only to offer clarity for those who do not know the primary term. This mirrors Ojibwe writer Renee Senogles's observations about how Red Lake Nation youth created a tribally centered interpretive trail along Pike Creek, where metal signs using images and text might say "Migizi" at the top, and then "way at the bottom *underneath in parenthesis*, it would say (eagle)" in order to also signal the sentiment that "this is who we are, we are Ojibwe" (Senogles 1996, 179; emphasis added). While the physical signs use both languages to meet oft-cited safety requirements, the Dresslerville tribal community reversed the usual bilingual ordering, in which English remains situated "on top," a physical and symbolic position that would normally indicate the direction of translation moving from English to the tribal language. Here, Washiw is centered.

I should briefly note that these observations should not be read as a critique of any tribal street sign strategies. All of the signage systems I discuss in this chapter represent different approaches, each of which offers differing possible consequences and interpretations. Further, desired outcomes are never guaranteed, and intentions frequently do not manifest in the ways one might hope or suspect. The more significant point is the fact of parallel spatial practices in diverse tribal communities from different regions of the United States as well as the community intentions, regardless of outcome. For all these Native communities, street names and signs mirror Washiw tribal elder Steven James's assertion about the Washiw, that "the people, culture and the language cannot be separated, the language is the identity of the . . . People" (Washoe Cultural Preservation Office n.d.). I would add only that these interconnections are not just about culture and identity, but also about space. Featuring indigenous constructions of the world on street names and signs in the form of tribal language and in the form of cultural content (as in Lame Deer) expresses an emphasis on tribal perspectives (culture, language, and politics) and physically marks the creation, reclamation, reassertion, and re-creation of indigenous geographies. Just as the independent proliferation of Indian-themed street names in White communities illustrates a larger racialized and cultural sharing as it plays out in making and giving

meaning to space, the use of tribal language and culture-based street sign systems indicates a deliberate response to the connection between tribal identities,
languages, cultures, and their land base. The use of such street names in Native
communities, then, might well be considered parallel to indigenous efforts to
use or reclaim traditional place-names, which often contain important cultural
knowledge and mechanisms for maintaining tribal identity within the names
themselves (Basso 1996). These efforts offer examples of productive actions
aimed at maintaining and restoring language and culture, reasserting identities,
and sustaining healthy relationships with the homelands. In short, posting street
signs with indigenous-language words or more direct representations of tribally
specific culture and practice is an act of community health and directly practices
(produces) tribal geography.

Other Spatial Markers

While my chosen focus for this chapter and the next has been street names and
signs because of the way they serve as concrete spatial markers in discrete yet
powerful ways, other elements of the larger sign systems are also valuable sites for
analysis. I want to highlight a couple of these related elements, derived from two
already discussed sites, Musqueam and Lame Deer. In both cases, other spatial
markers—a tribal logo and license plates—are intimately connected to the street
names and physical road signs. The logo, used in a number of contexts, is also
prominently placed on many of the reserve's street signs. License plates, on the
other hand, are legible in direct relationship to the reservation, its streets, and
the tribal-member-owned vehicles used to navigate these and other streets. Each
of these spatial markers, then, also extend the kind of work being done by street
names and signs. They stake out geographic claims and demonstrate consciousness over the importance of practicing and announcing tribal spatialities.

Like many municipalities, the Musqueam Nation includes an official logo
on their street signs. Although they appear as part of the signs, these logos also
extend the discussion of marking Native space beyond the street names and signs.
They illustrate how official spatial markers can overlap and intertwine with other
cultural practices and representational strategies. Designed by Musqueam artist
Susan Point, the logo depicts a salmon falling into a traditional fishing net, all
of which takes place inside a stylized, turquoise arrowhead (Musqueam Indian
Band 2006, 37). This is the same logo that flies on the Musqueam Nation flag,

FIGURE I.8 Signs at intersection of Thellaiwhaltun Avenue
and Crown Street on the Musqueam Indian Reserve, Vancouver BC.
Photo courtesy Harley Augustino.

and thus its inclusion on the street signs provides an additional pronouncement
of First Nations sovereignty and represents a conscious tool for (re)producing
and officially proclaiming Native space.

If one follows Crown Street from the intersection with Thellaiwhaltun Avenue
(fig. I.8) away from the reserve, we expectedly find that the Vancouver city signs
do not display the Musqueam logo. They become the standard, unadorned signs
found throughout the city. Paying attention to this simple difference reminds us
of the important distinctions between these two buttressed and related spaces
(Native and non-Native), but also points to the Musqueam Band's interesting
semiotic reinscription of the crown. In 1914, Crown Street was extended south-
ward from the Jericho naval reserve to the Musqueam Reserve boundaries,
a move that Snyders and O'Rourke call "a potent reminder to the tribe as to
who called the shots" (Snyders and O'Rourke 2001, 76). On the contemporary
Musqueam sign, however, the crown as ultimate colonizing symbol and authority is now simultaneously stamped with this marking of tribal sovereignty and
Musqueam space.

Such reinscription is neither purely discursive nor so fully disempowered that it no longer matters. In June of 2014, the Vancouver City Council passed a motion officially recognizing that the municipality was founded on and still inhabited the lands of three First Nations communities. The motion indicated that the council would now "formally acknowledge that the city of Vancouver is on the unceded traditional territory of the Musqueam, Squamish and Tsleil-Waututh First Nations" and craft "appropriate protocols . . . to use in conducting City business" (Vancouver City Council 2014). The proposed interventions were created to "respect the traditions of welcome, blessing, and acknowledgement of the territory" even if they appear to be somewhat vague, limited to ceremonial greetings and opening prayers and thus easily achievable for the municipality. This formal stance undoubtedly signals a major shift in cultural, philosophical, and governing principles in a settler colony city, with the potential for further enhancement in future efforts toward "reconciliation." For my purposes, it is also impossible to ignore that the motion was tied to the Standing Committee on Planning, Transportation and Environment, an entity explicitly charged with duties related to land, land use, and human relationships to the land—in many ways, issues of geography and spatiality.

While the discursive and iconographic play in Musqueam between Crown Street and Thellaiwhaltun Avenue is probably a happy coincidence rather than a result of intentional critique, the intersection (both the literal and figurative) nevertheless symbolizes the band's larger spatial overlaps with Vancouver and the Canadian nation—a geographic and now potentially productive tension being formally embraced and recognized by the Vancouver City Council. In their 2007 Musqueam Community Profile, the band articulates a desire for the larger Vancouver community to reinsert tribal place-names onto the landscape (Musqueam Indian Band 2007). This expectation formally extends beyond the reserve, and thereby looks to institutionalize acknowledgment of the wider traditional homelands on which the city is built. Musqueam, in effect, seeks to expand its spatial marking campaign across their traditional lands, which means challenging the presumed-exclusive, non-Native geography of Vancouver.

In a move parallel to Musqueam's use of the tribal logo on street signs, we might consider the Northern Cheyenne introduction of vehicle plates. Adopting an official license plate (see fig. 1.9), the Northern Cheyenne joined a number of Native nations across the United States that have done so since the Red Lake

Chippewa in 1974 (Tribal Council of the Northern Cheyenne Tribe 2010). The explicit spatial element of this practice is not insubstantial, given that one of the primary definitions of a nation or state is control over an established land base and the mechanisms for defining the relationship between residents/citizens and that land. License plates serve as physical reminders of a defined territory and its correlated cultural and political identity. They locate travelling bodies and objects (vehicles) and "place" them, or provide them a static home/land. Besides identifying the vehicles and driver, license plates signal jurisdiction and the legal relationship between person, vehicle, and the spaces through which they travel or come to rest. They indicate the proper place of one's inhabitation and mark one's inhabitations as proper. They operate as moving signs on the streets. Taken together, then, license plates and street signs operate as mobile components of a multifaceted sign system—again, Edensor's "signscape": a productive arrangement that informs travellers, helps produce subjectivities, and both marks and crafts particular geographies (Edensor 2004).

The significance of such a political statement is not lost on state authorities either, as demonstrated by a number of recent court cases. In Kansas, authorities questioned the legality of Potawatomi-issued plates. The federal courts eventually sided with the Prairie Band of Potawatomi in 2007, rejecting the state police's ability to discriminate against tribal governance (as opposed to other states) and arrest or cite tribal members for driving with "illegal" non-state-sanctioned plates. In other cases, legal limits placed on non-recognized tribal license distribution were actively flouted. Geographer Jonathan Leib points out, for example, that the Abenaki in Maine intentionally issued non-recognized tribal license plates precisely to represent tribal sovereignty while they were seeking state and federal recognition (Leib 2011, 48).

As with street names and signs, of course, license plates also have the potential for "locating" and controlling Native peoples and communities. This is indeed one of the primary functions of any license. Yet, many tribes have likewise found it more prudent to take on such bureaucratic mechanisms. This is motivated by a combination of factors, such as fear of outside control or economic incentives, but also clearly illustrates an active expression of spatial sovereignty within the confines of mainstream legal mandates. Consider how the Cherokee Nation and the state of Oklahoma agreed on the issuance of statewide-recognized tribal plates, signing a compact in 2013 and officially registering the tribal plates with the state's vehicle database. These brief examples point to the ways that states have an interest in the production and usage of something as seemingly mundane as license plates and street names and signs precisely because of their symbolic, economic, and political meanings. The installation of signs and the issuing of license plates assert and confirm that the current administrative power of tribes is partly rooted in traditionally based geographies that refuse their political or spatial erasure despite the insistence of some state or federal bodies.

We should also quickly note how the design of the license plates is deeply meaningful in terms of culture and spatiality. The Northern Cheyenne plates use a traditional color palette of red, yellow, white, and black, representing the four sacred colors and the four directions, and commonly used for medicine wheels. Two of the identifying texts—"Montana" and "The Northern Cheyenne Nation"—are highly stylized in the distressed and flourished "Bleeding Cowboy" font (Last Soundtrack 2007). In a much more sober serif font at the bottom of the design we find a traditional self-identifying statement: "The Tsististas & So'taa'eo'o People." Grounding the entire scene, the background silhouettes Noavose, the Sacred Mountain or "The Hill Where the People Are Taught," a place where the Creator revealed himself to the prophet Sweet Medicine (Ambler et al. 2008, 85), and origin of all Sacred Powers (Powell 1979, xxxvii).

On either side of the mountain are representations of the Northern Cheyenne's most revered historic figures, Little Wolf (or Little Coyote) and Dull Knife (or Morning Star), rendered from a famous photo of these men on their 1873 visit to Washington, DC, where they were protesting against confinement to Indian Territory. They travelled a great distance to represent their people, to articulate tribal land rights, and request the opportunity to return to their northern homelands (Powell 1979, 832). Their frozen image from that moment is now iconic in the Northern Cheyenne communities. It effectively captures the geographic desires of the people and generates an ongoing source of their identity as willing to risk all for their relationship to the homelands. Today, all tribal council meetings are filed on documents watermarked with this photograph and the statement "Little Wolf and Morning Star—Out of defeat and exile they led us back to Montana and won our Cheyenne homeland that we will keep forever." This same image is also found inside the official "Great Seal of the Northern Cheyenne Tribe," which is situated in the bottom left corner of the tribe's license plate, and thus visually doubled on the license plate.

The design and collage of images on the Northern Cheyenne tribal license plates captures an important material representation of Northern Cheyenne

FIGURE 1.9   Northern Cheyenne license plate design.
Courtesy Montana State Motor Vehicle Division.

spatiality. Beyond their administrative and economic functions, these images operate as cultural documents referencing this community's daring negotiation of exile and return. They reflect and reinforce contested claims on a homeland. Much like the rhetorical and discursive work done by placing Thellaiwhaltun Avenue against Crown Street in Musqueam, the license plates can extend the reach of Northern Cheyenne geography, as the cars bearing such plates travel through other lands and draw attention by other motorists, pedestrians, state officials, or law enforcement agencies.

Despite the inherent tensions and potential failures in practicing sovereignty, control over spatial tools seems the best option for tribes. The enactment of control over and the material expressions of indigenous geographies mirror other regained aspects of self-governance and self-determination: language, education, health, and economic development. These choices may be deliberated over with great care and intentionality, or simply enacted with bureaucratic detachment. The ultimate decisions nevertheless fit into a larger network of decisions and practices that work to produce particular kinds of space. Those networks of decisions are related to and connected with other decisions about everything from repatriation to battles over mascots, from ceremonial gatherings to gaming policies. While street naming may not be understood to be as important as these other decisions about tribal life, street names and signs are understood in many communities to have value, and are certainly more directly and explicitly about space than most any other aspect of tribal life except land claims.

Perhaps we can expect the differences we find in the use of indigeneity/ Indianness in sign systems and spatial marking practices between Native and non-Native communities (which I explore further in the next chapter). As I have already suggested, however, instructive differences and similarities are also found between Native communities. In thinking through differing approaches to street naming in making Native space, we can turn to the names and signs chosen by these sites to concretely illustrate the variety of approaches, whether found in Musqueam or Dresslerville, or elsewhere. These choices are undeniably community-based and tribally centered. At minimum, this delineates key differences between street names in tribal communities from those in non-Native communities. Looking at additional markers such as the logo and the license plates, we can also point to the larger realm of spatiality and the possibility of tracking other modes of recognizing, expressing, and challenging space, especially those with overlapping tensions.

2 Inhabiting Indianness in White Communities

Three Visits: Chepenafa Springs, Dunawi Creek, Tualatin

I am standing in front of a sandbox full of community toys. I can see two commercial play structures in green, red, and yellow. Carefully watered and trimmed crabgrass. We are well above much of the city here, but the abundant trees outlining this park obscure my view. Chepenafa Springs is a city-owned park along the relatively well-to-do northern edge of Corvallis, Oregon. It occurs to me that the name of this park likewise rises above the standard American use of Indianness in place-naming. Until recently, names that were not historic carryovers from tribal languages rarely found their way on to local places. Here, one has. Looking through city council minutes, I find that the name was rather casually chosen to acknowledge the Chepenafa, or Marys River Band of Kalapuya, who lived at the confluence of the Marys and Willamette Rivers, precisely where Corvallis was founded in the 1840s (incorporated 1857). I desperately wish there was more to the choice. Nothing at the park explains this choice. This "elevated" naming still reflects a limited vision for reconceptualizing Native and non-Native relations and histories—for addressing questions of indigenous and occupied space. Chepenafa Springs is abstract.

By 1859, the federal government had removed the Chepenafa and all of the other Native people in the surrounding areas to the newly established Grand Ronde and Siletz Reservations. I look around from my vantage point and wonder what kind of work this park's name can do (or undo) in relation to that legacy? I am mostly skeptical that there is any real expectation of work. Maybe just a nod toward inclusion. Rhetorical multiculturalism. I acknowledge that this is a lot to ask of a name, of a park where people go to enjoy themselves and reduce or escape the stresses of their daily lives. But why else name the park this way unless there is an intention to make a statement, to concretely proclaim a set of values regarding the indigenous peoples of this land?

This park's name is not the only statement of this kind made by the City of Corvallis and its Parks and Recreation Department. A few months later, I walk on a recycled plastic boardwalk reading informational signs posted along Dunawi Creek. *Dunawi* is said to translate as "wise woman." In 2000, at the urging of a local college student, this small waterway was renamed to replace its previous Squ*w Creek designation—again, dis/spelled here following C. Richard King's example (King 2003). When I read the brochure describing this slice of restored wetlands sustained by the creek (and its resident beavers), I learn that the name change is narrated as part of a "nationwide effort to remove [that] derogatory term . . . from place names" (Benton Soil and Water Conservation District 2013). I am impressed by the political agency and how the environmental reconstruction efforts fit into the larger embrace of wetlands as important ecosystem niches. In this case, the restoration is also solidly couched in its replanting of tribally important plants. I read that "the planting design . . . included vegetation used in Kalapuya basketmaking." Admirable, ecologically and culturally. Except that, so far as I can determine, no Kalapuyas or other tribal people gather materials here. In fact, the boardwalk where I walk is intentionally constructed for "human visitors and their pets" to (paradoxically) leave the "ecosystem undamaged while they explore the wetlands." In other words, I am treated to the ecological parallel of "stay off the grass" signs that usually protect landscaping aesthetics.

The precolonial landscape reflected knowledgeable and skilled human intervention. The plants need it, just as they need the beavers. Responsible exploring. The source and applied laboratory of traditional ecological knowledge. Here I am offered isolation and preservation. Resisting the protectionist message, I hear Potawatomi ecologist Robin Kimmerer's voice in my head: the plants' significance, their very health, relies on a particular relationship with people, not just the plants' restored and managed existence (Kimmerer 2013). Otherwise, preservation abstracted. Indianness abstracted. So, I struggle to find the value of including a name or plants or referencing a Native geography and botany if the land and the people named remain colonized and alienated from each other. Yes, it illustrates Renato Rosaldo's notion of imperialist nostalgia. Yes, it undoubtedly assuages White guilt. I do not deny these are partially valuable sentiments. But to become meaningful, such sentiments must be paired with substantive relationships. Like the plants, anticolonial relationships will not flourish in the abstract. So, I worry that we too often expect such good intentions to impossibly reconcile material consequences through language alone. Where do words go, and what do we do with them?

Speaking of abstraction: Before a trip to the local bowling alley to meet with a colleague, I notice how the town of Tualatin, Oregon, like all suburbs, presents spatial abstraction via neatly ordered residential enclaves and parks. Curious, I look up the town's name. Tualatin comes from an unidentified and thus abstracted "Indian word impossibly translating to anything from lazy to sluggish to treeless plain to forks to the name of the local Kalapuya band, the Atfalati" (McArthur 2003). The street names here are also abstractions of Native peoples. Standing on a street corner, I look up and see Sioux Court on a green and white sign embellished by the town's vaguely tribal-inspired "T" logo. Apache Drive. Cheyenne Way. Iroquois Drive. Piute Court. Chinook Street. All bunched together. All out of context. All the same. Neatly organized. Indians. Just like in Clairemont Mesa, California. Just like Ahwatukee, Arizona. Just like Cherokee Village, Arkansas. Just like Medford Lakes, New Jersey. And so on. All of these towns have Cheyenne Avenue, Drive, Trail, or Way. They all have Sioux, and Iroquois. Apache, Chinook, and Piute/Paiute. Three of them host Tonto, either referencing the so-called Tonto Apache or the most famous fictional Indian of all. At least that fiction would be honest, upfront. If only Natives were not also mythologized, that might provide a good start.

A Nation of Indian Villages

During my youth in Santa Rosa, California, I was always curious about a part of town referred to as "Indian Village." I did not spend a great deal of time there, but I knew that it partly mirrored my own neighborhood in the Roseland area on the other side of town, which is to say it was a low-income and relatively racially and culturally diverse pocket within a mostly White town. Both areas were increasingly marked as disreputable, and publicly identified as such by their higher rates of crime, gangs, and street violence. As a young person I grappled with the discourses about these spaces, and the rampant racial and cultural implications that implicated me and my friends, neighbors, and families.

To make sense of these spaces and discourses, I spent a lot of energy thinking and learning about how structures of diversity, inequality, and racism shaped these spaces located within but seemingly distanced from the surrounding rural-suburban landscape. I also developed a lingering but only later pursued interest in the name Indian Village. Having lived on a reservation, and grown up along the edges of urban Native communities there and elsewhere, I was certain that Indian Village was not a Native space. That section of town was not home to many American Indian people, or at least no more so than other parts of town. It certainly was not a retained or even reclaimed tribal territory, as those both existed nearby. Rather, the Indian Village moniker merely recognized that this collection of residential streets was named after Native peoples. Navajo Street, Cherokee Avenue, Pomo Trail, Cree Court, Sioux Drive, Seneca Lane, and more.

Only later in life would I start to unpack my lingering interests in this place-name and begin to see how the discourse about Indian Village actually worked quite well with long-standing constructions of Indianness and Native space as violent. Symbolically rendering this neighborhood as a throwback to the "wild west" or to precontact "savagery" meshed well with the "urban" challenges of poverty, violence, and racial tensions. Both marked a supposedly nonrational space containing difficult, racialized citizen-subjects seemingly intent on disrupting an otherwise peaceful community.

Although Indian Village contained the discursive potential to revisit the Savage trope, it is also clear that home builders' deployment of Indianness as marker of residential space intended to reference that trope's Noble Savage counterpart. Using this image, the developers could identify with and perhaps even eagerly want to *inhabit* Indianness. Far from marking an undesirable space, the intent seems to have been to mark it as an eminently desirable location for purchasing a home and raising a family.

As a recent phenomenon, Indian-themed street naming projects found all across the national landscape work in this way to recover an Indian figure through the more intentional goal of producing attractive and pleasing domestic environs and ultimately garnering profits from home sales and equity wealth. As Philip Deloria notes, appropriations of Indianness have continually been deployed toward various goals and social projects for modern, Western society (Deloria 1998). In its more "positive" usage, the Noble Savage-as-figure represents a salvageable racial subject recovered by and for modern citizenship. As scholars have long argued, these two models of the Savage and the Noble Savage work in tandem and have been deployed as heuristic tools by which to praise or critique Western cultures and societies (Berkhofer 1978; Bird 1996; Deloria 1998; Huhndorf 2001; Mills 1997; Moses 1999; Said 1979; Strong 2013). With both models, however, actual Native peoples are less important than the reflective, intellectual exercise provided by the figure of the Indian; the Other serves as simply mirror or counterimage for the construction of the Western Self of modernity (Bhabha 1990, 1994; Deloria 1998; Lott 1993).

Coinciding with the ongoing cultural role of the Indian, such deployments of Indianness also rest on the related framing of Native space as a utopic, premodern space of freedom. This understanding of the relationship between Indianness, the construction of physical places, and the production of space finally came together for me while living in another city. Working on a cultural studies project for a graduate course on race and media, I was drawn to one particular street sign in San Diego, "Manito Way," referencing an Ojibwe/Algonquian word roughly meaning "spirit" (as narrated in chapter 1). Drawing on my until-then-submerged interest in Indian Village as a peculiar set of spatial markers, I located this street on a map and quickly saw that housing developers had branded an entire residential area with Indian-themed street names. Another Indian Village, it seemed. My youthful curiosity suddenly crashed into a set of research questions about the intersection of Indianness with space-making practices, Whiteness, indigeneity, and nationhood. Through my research I ultimately uncovered a vast network of such "villages" across the nation. While Santa Rosa's Indian Village was locally known as a space of color and of danger, the vast majority of the Indian Villages I found proved far less diverse—and far more idyllic in space- and meaning-making practices. Their demographic makeup proved conspicuously White, suggesting that residential spaces using Indianness as a set of spatial markers were designed through a White lens and as intentionally White spaces. In these residential areas (and they are always residential), Indianness worked toward narratives of a domestic and mythic ancestry, and thereby in support of a racialized national unity, democracy, and (ironically) multiculturalism.

In thinking through the process of individual interpellation into settler colonial spatialities, Mark Rifkin asks, "When and how do projects of elimination and replacement become geographies of everyday non-Native occupancy that do not understand themselves as predicated on colonial occupation or on a history of settler-Indigenous relation (even though they are), and what are the contours and effects of such experiences of inhabitance and belonging?" (Althusser 2001; Rifkin 2013, 324). While Rifkin ultimately argues that settler common sense is ostensibly contained within everyday activities and understandings that have no direct reference to Native peoples, I contend that the depth of normalization can be such that Indianness can be invoked and infused without concern for any self-reflective consideration of colonial occupation. Experienced solely as an aesthetic encounter, residing on lands textually referencing Indianness remains likewise abstracted from legal, political acts of dispossession and settler occupation. Precisely because of the mundane nature of street signs and names, Indianness can be both directly referenced and removed and thus serve as the "quotidian" means by which "settler sovereignty is activated, circulated, and materialized" (Rifkin 2013, 326). Street signs and names can signal (and further help remake) settlement without directing attention to the histories or official structures of colonization, thereby routinely and simultaneously possessing Indianness and dismissing indigeneity. Residents simply purchases homes and live their daily lives, and thus casually enact the continuity of settler space.

Crafting White Space

At least as early as 1900, American city planners marked groups of streets using an Indian theme. In that year, Philadelphia's Chestnut Hill neighborhood changed a series of numbered streets—Twenty-Seventh, Twenty-Ninth, Thirtieth, Thirty-Third, and Thirty-Fifth—to Shawnee, Navahoe, Seminole, Huron, and Cherokee (Alotta 1975). In 1912, the Atchison, Topeka and Santa Fe Railway company created out of whole cloth a notable cluster of Indian-themed streets. This cluster comprised every street in the agricultural town of Satanta, Kansas (which I treat in chapter 3), and epitomizes the more intentional, comprehensive, and market-interested Indian-themed sign systems that became widespread between the 1950s and 1980s.

The Indian-themed street name cluster phenomenon marks the intersection of a moment of concentrated residential construction (the "housing boom") with a "national pastime" of appropriating Indianness (Deloria 1998). Consider Cherokee Village, Arkansas, founded in 1954. Since its origin, this recreational and retirement community has grown to encompass fifteen thousand acres, wherein all the lakes, parks, and community structures draw on the Indian theme. Residents traverse no less than 150 individually named streets, each of which draws on a unique combination of abstracted or Cherokee-inspired names, such as War Eagle, Cochiti, Blackfoot, Sequoyah Ridge, and Talequah.

In 1956, developers for a new residential area of South Lake Tahoe decided on Apache and Hopi Avenues as two of their earliest street names. On their way to producing another of the largest Indian-themed clusters in the nation (see fig. 2.1.), the builders added locally relevant names like Piute and Washoe (in 1958), and then more nationally scaled names like Kiowa and Seneca (in 1961) to its growing roster of streets. By 1969, all or most of the one-hundred-plus current streets were in place, with residential plots strategically available for the coming real estate and population growth.

The relatively narrow time frame during which such communities were built captures a period of conscious and intentional effort by developers and marketers to represent Indianness as a key component of their economically driven projects. They provided a new mode by which Americans could insert Native ghosts to help narrate a newly designed landscape (Bergland 2000). Indianness worked to craft attractive new domestic spaces, places for real inhabitation. Building on a long and successful tradition of commercialized Indianness and product-branding techniques, builders clearly understood that Indian themes were marketable to would-be homeowners. They indirectly referenced nature. They offered romantic mythologies. They were "native" to the land of this nation. They were original. They conferred history and tradition to newly made residential spaces without either.

Before saying more about the unique use of Indianness in these commercial-residential-spatial projects, I want to briefly acknowledge that many place-names carry lineage from Native vocabularies. Those names have been either actively or unwittingly passed along into everyday and modern usage. For the purposes of this chapter, however, I am less interested in the undeniable fact that numerous places such as Miami, Lake Tahoe, Oklahoma, and the Walla Walla River bear names rooted in Native-language origins. For this, we can look at William Bright's *Native American Placenames of the United States* (University of Oklahoma Press 2004), or the two-volume set by Sandy Nestor, *Indian Placenames in America* (McFarland 2003).

I am interested in the fact that mid- to late-twentieth-century developers with substantial commercial interests consciously tagged residential streets across the nation with names like Apache, Cherokee, and Tomahawk, and then packaged these names together into thematic spaces that abstracted American Indian cultures, geographies, and histories. It is also vital to note that such promotions found a receptive market. People bought the houses. They participated in more or less active ways in using these articulations of Indianness in one of the most intimate parts of their daily lives.

FIGURE 2.1 Detail map of Cherokee Village, Arkansas, the nation's largest Indian-themed street name cluster. Courtesy © OpenStreetMap contributors.

Via such developments, Indianness-as-street names became uniquely prolific. Why unique? What do I mean by prolific? Let me start with their proliferation. We find them in all sorts of communities. These communities are in urban and rural settings, in communities from every region of the country. Few states lack such clusters. Most hold a number of substantial examples. The vast majority are made up of tribal names: Choctaw, Sioux, Arapaho, Delaware, Chinook. In brief, Indian-themed street sign systems are present across the national landscape and surprisingly uniform in articulation. Compare the suburbs of Ahwatukee (Phoenix, Arizona) and Tualatin (Portland, Oregon) against the rural, recreational communities of Cherokee Village (Arkansas) and Lake Tansi (Tennessee). Phoenix is the nation's fifth-largest city, with more than 1.5 million residents. Tualatin is part of Portland's two-million-plus metropolitan population. On the other side of the spectrum, the community of Cherokee Village is home to fewer than five thousand permanent residents; Lake Tansi, fewer than four thousand. Taken together, these sites succinctly illustrate the range of such clusters across community size, type, and region. So, they are prolific to the point of being mundane. Most of us have seen them, travelled them, ignored them. Many people have happily or guiltily confided to me that they or their families have lived on one of these streets.

With such scale at work, my previous scholarship grouped clusters of Indian-themed street names into five categories: tiny (up to ten streets), small (eleven to twenty streets), medium (twenty-one to forty streets), large (forty-one to ninety-nine streets), and super (one hundred or more streets) (Barnd 2010). Using this ordering, I produced a low-end estimate of 180 distinct sites across the country. Only a small handful of these towns and cities exclusively feature such names, as found in Satanta and Cherokee Village. Most resemble South Lake Tahoe's cluster: a group of street names ultimately hidden by the totality of the community's many other streets. So, Indian-themed street name clusters tend to be obscured by both the general banality of street names as well as by the larger pool of names comprising the town street names. Flouting this invisibility, however, they collectively encompass more than ten thousand individual streets of American residential space.

So, what makes this practice unique? In short: Whiteness. As table 2.1 highlights, the residents of the largest Indian-themed street name clusters (four of which are presented here for their demographically representational value) averaged 89.7 percent White in 2010, down slightly from 91 percent in 2000. These

TABLE 2.1. Percentage of White residents in selected Indian-themed super clusters (100 or more street names), 1980–2010. Source: US Census.

Community	% White 1980	% White 1990	% White 2000	% White 2010
Cherokee Village, Arkansas	99.3	99.1	98.7	97.2
South Lake Tahoe, California*	n/a	95.1	91.3	92.2
Ahwatukee, Arizona*	n/a	94.2	88.8	80.7
Tualatin, Oregon*	95.1	95.8	89.3	82.9

* Totals for these towns are aggregated using only those census tracts composed of Indian-themed streets, and excluding all others. Census tracts shifted between years and sometimes included additional areas immediately surrounding the targeted streets.

n/a: Census tracts were not yet used for most smaller areas before the 1990 census.

statistics can be compared with the White population in the United States as a whole, which likewise declined, but dropped from 75.1 percent in 2000 down to 72.4 percent in 2010. Thus, the largest cluster sites, on average, house a 17 percent greater proportion of White residents than the nation as a whole, and have decreased their White population percentage at a slower rate. Further, many of these sites house a relatively higher rate of White residents than even their immediately surrounding communities. Consider that the broader local demographics of such sites only average 87.9 percent White residents. This means that even in relation to the most local context, the Indian-themed clusters still prove "Whiter," demographically speaking. As the early examples of Philadelphia and Satanta hint, and these more contemporary statistics confirm, the usage of Indianness for residential street designations is racially coded as and for White space. In some instances (such as in Clairemont Mesa, California), we can trace how, historically, they have been specifically produced directly via restrictive residential practices and policies.

What also makes this correlation between Indian-themed street names and White residents unique is that the racial coding reverses what we normally find with commemorative and place-naming practices. Formal spatial markers used to reference this particular set of racialized (and importantly, indigenous) peoples identifies places controlled by and "belonging" to the dominant nonracialized population. In other words, the Indianness applied to places like Philadelphia and Satanta marks residential clusters historically inhabited by White occupants.

This correlative spatial relationship suggests that searching for normatively constructed Indian-themed street names—in contrast to those tribal examples highlighted in the previous chapter—might serve as an unusually effective if indirect method for locating a heavily White demographic.

To further mark the significance of this observation, let's briefly consider other street names referencing racialized peoples. Highly visible political battles have been and continue to be fought over street names (usually renamings) that are explicitly sexualized or racialized, such as Martin Luther King Jr. boulevards (Alderman 2000, 2003; Alderman and Inwood 2013; Mitchelson, Alderman, and Popke 2007). Most White Americans are comfortable with the idea of Martin Luther King Jr. as an American civil rights figure, although few seem comfortable with being associated with (presumed Black) spaces that are marked by his name. As Derek Alderman and colleagues have documented, White Americans resist naming "their" streets after African Americans for fear of the presumed economic and social repercussions of being identified as a Black space. In the discursive intersection between race and space, Martin Luther King Jr. streets mark locations commonly accepted as Black geographies, just as the ever-increasing César Chavez–named streets always indicate a significant Latino population and cultural presence. Such public namings, however, are confined gestures to be applied as nominal replacements for previously "neutral" street designators. So, Main Street becomes MLK when demographics shift. This comparison also helps demonstrate that commemorative-based themes celebrate individuals (usually historic) rather than explicitly representing group identities (Black, African, Ashante) or making cultural references. To some extent this explains why, despite what "common sense" might suggest, American city builders have produced absolutely no significant street-name clusters referencing any other racialized group besides Native peoples, let alone women or nondominant sexual identities. Political representatives choose only single street name replacements rather than immensely more complex thematic changes.

This distinction in street name content is accompanied by a distinction in quantity. Streets named after Martin Luther King Jr. certainly offer the largest aggregate of a single street name referencing a racialized population, remembering that references to White individuals or identities are deracialized as normative. Yet such hard-earned memorial naming is isolated to the degree that the approximately eight hundred King-named streets and boulevards are rarely clustered together with additional African American–themed street names. In contrast, Indian-themed clusters—which regularly encompass from 40 to 150 streets each and combine to a nationwide grand total exceeding ten thousand streets—dwarf the few African American–themed clusters that exist (Barnd 2010). In short, Indian-themed street names are abundant, frequently lumped into significant clusters, minimize individuals, and consistently mark spaces that were intended for and remain predominantly occupied by White residents. In all these ways, Indian-themed street names are quietly apolitical and only implicitly racialized because they operate as articulations of Whiteness and of White space. Spaces crafted through Indianness are widely acceptable in ways that those crafted by Blackness are not.

Closely connected to the relative Whiteness, these communities are also correspondingly absent a Native demographic presence. None of the largest twenty-six Indian-themed street name clusters house any substantial American Indian population. All around Ahwatukee, Arizona, we actually find substantial Native communities; four reservations are nearby (Fort McDowell Yavapai, Ak Chin, Gila River, and Salt River Pima-Maricopa), as well as a large urban Native community in the Phoenix area. Yet this Ahwatukee community is exceedingly White. Arizona, which ranks seventh in American Indian population percentage (reporting 4.5 percent), provides the only example of a state with a significant and nearby American Indian demographic presence and a large or super-sized Indian-themed cluster. Although separated from Phoenix proper by a mountain range, Ahwatukee was incorporated (piece by piece) as one of the city's "local villages" between 1978 and 1987. In 1990, the four census tracts that constitute this super cluster suburb still reported a 94.5 percent aggregate of White residents (tract 1,167.07 at 92.9 percent; 1,167.08 at 96.3 percent; 1,167.10 at 94.1 percent; and 1,167.11 at 93.9 percent), compared with the city's overall 81.7 percent.

Other states with relatively large American Indian populations (Alaska, 13.1 percent; Montana, 6.3 percent; New Mexico, 9.7 percent; North Dakota, 5.2 percent; Oklahoma, 6.8 percent; South Dakota, 8.6 percent) host only a few major clusters. None of these states are found on the list of twenty-six largest clusters. In such spaces, inhabiting Indianness likely becomes a much more clearly contestable and concrete practice. This reiterates the availability of the Indian for commercial and symbolic usage, and its construction largely outside of the purview of Native peoples and communities, and suggests that when that purview is opened to or just geographically near substantial Native populations, the practice loses momentum.

It should be noted that, historically, urban clusters increase their populations of color slowly over the decades after inception, although usually only after White residents leave for suburbs and other "Whiter pastures" (although we see this trend shifting now with urban gentrification). In June 2001, the demographic changes in Ahwatukee were finally noticeable enough for some residents to feel encouraged by the changing face of what had long been dubbed "All-White-Tukee" (Biggs 2001). The White population percentage continued its downward trend and had dipped to 80.7 percent by the time of the 2010 Census, although that number still significantly trails the shifts occurring in Phoenix as a whole, where the White demographic makes up just 65.9 percent of the population, with only 46.5 percent claiming a White, non-Hispanic identity.

In spaces less urban than San Diego (home to the Clairemont Mesa cluster) or Phoenix, such White flight and demographic shifts may never occur. This allows those communities to retain their stable and mostly White demographics. Cherokee Village, Arkansas, attests to this stability. Arkansas hosts no federally recognized reservations. Almost no self-identified Native individuals live in Cherokee Village or the surrounding region. And as table 2.1 indicates, it remains more than 97 percent White, representing just a 2 percent drop in thirty years.

The Work of Inhabiting

Neighborhoods with extensive Indian-themed street names differ from one another in density, design, size, and location, but all share a demographic reality shaped by the relatively extreme Whiteness of the residents and the lack of Native peoples. As mentioned, this use of Indianness as spatial marker extends across state and regional differences. Thus, Indian-themed street name clusters locate a conceptual, demographic, and geographic Whiteness. As is true of its many popular culture uses, the Indian crafted and deployed as street names for non-Native communities is by definition circumscribed precisely by its non-Native possession. In some ways this is fitting, given that Europeans invented the Indian (Berkhofer 1978). Also fitting, then, we see that the street name content is standardized and abstracted. As I noted earlier, most are tagged with tribal names. The rest are compiled from a small handful of cultural items ("teepee"), historic figures (Tecumseh), fictional characters (Tonto), and approximations or simulations of tribal languages (Chief Big Look) (Barnd 2010). The particular construction of Indianness, and the logics that inform how Indianness can be expressed in spatial terms, tells us a great deal about its relation to Whiteness. Both Whiteness and Indianness have predetermined geographies that are made further apparent when surveyed through space-marking street sign systems.

Looking over the maps with me, colleagues are often struck by the fact that Indian-themed clusters are frequently located near golf courses, bodies of water (usually lakes), and other (sometimes fabricated) idyllic settings. The streets constituting the neighboring communities of Enchanted Oaks/Payne Springs (Texas), straddle the thirty-two-thousand-acre Cedar Creek Reservoir. The mobile home park cluster in Fort Myers, Florida, is situated mere minutes from the Gulf of Mexico. Most of the large and super clusters provide excellent examples of this practice, which are often clearly indicated by their very names, including Lake Havasu City (Nevada), South Lake Tahoe (California), Country Lake Estates (New Jersey), and Lake Royale (North Carolina). Cherokee Village boasts no less than seven lakes, each likewise following the community theme: Aztec, Chanute, Cherokee (of course), Navajo, Omaha, Sequoyah, and Thunderbird.

The tendency to associate Indians with natural and environmental features is doubly emphasized by the frequent use of the street-name qualifier "trail." Numerous street clusters are entirely constituted by streets with names like Iroquois Trail and Shawnee Trail. The roads of Country Lake Estates, Medford Lakes (New Jersey), and the Lake George region of Colorado epitomize this practice. Fort Myers Beach (Florida) offers fifteen parallel one-way "Trails." Although some Trails may coincidentally reference a historical relationship between the current street location and the traditional travel routes of Native peoples, most are purely decorative. The Indian-themed mobile home park in Fort Myers Beach, for instance, offers no Trails for the Seminole or Calusa nations from Florida. They do, however, include Trails for geographic outsiders like the Apache (from Arizona), Blackfoot (from Montana), and Seneca (from New York). This exemplifies a rhetorical Trail of Exclusions common in such spaces.

These spaces also reliably generate new, largely imagined links between Native nations without substantial geographic or historical connections. The community around Towamensing Lake, near Albrightsville, Pennsylvania, hosts a bevy of Trails that bring together such unlikely historic intersections as Chinook and Cochise, Piute and Narragansett. Historian Robert Alotta notes the same discrepancies in his local study of the early street renaming in Philadelphia. Tracing

FIGURE 2.2 Dakota Drive and Tomahawk Lane street signs in Clairemont Mesa (San Diego), California. Photo by author.

the street name selections and changes, he recognizes their abstraction in relation to the city and was left no choice but to argue that his city's Indian names must have been selected simply because "they sounded good," since there was such an extreme lack of either "local significance" or "geographic similarities" (Alotta 1975). What sounded good in Philadelphia actually echoes national, racialized appropriations of Indianness. Such conflations of Native nations are materialized on numerous intersections among tribal names, and among tribal names and "affiliated" object names like "teepee" or tomahawk (see fig. 2.2). Tomahawk, for example, is probably the most common object referenced and is found on no less than 612 streets, most of which are in Indian-themed street name clusters (Barnd 2008, 114). In all, the sign systems make persistent links between Native people and idyllic representations of "nature," and then sometimes compile them in such a way as to simultaneously exploit and displace local Native presence. Myths, after all, exist in the imaginary, and do not reside in the here and now.

Indianness and Whiteness

During the initial colonization of the Americas, the imposition of new names on the landscape reflected European need and desire for justifiable occupation of Native lands. Renaming and claiming territories delegitimized Native land rights and Native knowledge, as new European-derived names were deemed more appropriate for freshly "civilized" spaces. Many namings were explicit acts of appropriation and statements of precise land claims. Names were concomitantly chosen to extend European cultures and geographies symbolically (for example, New England, Cambridge, and Virginia). Even where such explicit Europeanizing of the named landscape was not present, however, colonial designations often belied implications and intent. New York City's Wall Street, for example, originated as a tool for protecting colonial space and actively excluding Native peoples while claiming the land from under their feet. Although the Dutch eventually abandoned their colonial post on the island when driven out by the English in 1664, their fortified, *walled* supply road remade a Native landscape into a space that discursively and physically protected invading settlers from Native inhabitants and thus marked Native peoples as dangerous trespassers on European lands.

In contrast to overtly colonial place-naming practices, mid- to late-twentieth-century Indian-themed streets offer an example of seemingly anticolonial naming practices. Whereas New Jersey was named after the British island of Jersey during the seventeenth century, the bucolic streets of Medford Lakes, New Jersey, are now replete with labels like Apache, Cheyenne, Mohawk, and Seminole (all qualified as "Trails"). Modern residential spaces like Medford Lakes reveal how twentieth-century housing developers have actively included notions of Indianness in their spatial projects. This offers a striking contrast to the early European colonists' need to rename, deny, and thereby claim Native spaces through discursive and physical markers of exclusion.

Despite such rhetorical changes, placing Indian names onto the landscape proves a disingenuous and ineffective "reversal" of colonial impositions over geography and epistemology. Such landscapes are most decisively not Native spaces. They remain thoroughly settled, colonized. Further, the towns and cities featuring Indian-themed odonyms (street names) are resolutely disconnected from Native communities, except that their very existence is directly linked to dispossession and the ongoing occupation of the Americas. As gestures of anti-conquest, these street names function as potent epistemological and onto-logical signs (literally and figuratively) that articulate and justify the material consequences of colonialism. Language and literature scholar Mary Louise Pratt uses the term "anti-conquest" to describe a postcolonial shift in "strategies of representation whereby European bourgeois subjects seek to secure their innocence in the same moment as they assert European hegemony" (Pratt 1992). Pratt uses this concept to describe travel writing, situating its discursive work newly invested in a "system of nature as a descriptive paradigm [that] was an utterly benign and abstract appropriation of the planet. Claiming no transformative power whatsoever, it differed sharply from overtly imperial articulations of conquest, conversion, territorial appropriation and enslavement. The system created ... a utopian, innocent vision of European global authority" (Pratt 1992, 38–39).

In the context of spaces that lack Native peoples, Indian-themed street names either locate Native peoples in a distant past or attempt to seamlessly incorporate them into a multicultural present, or both. In all cases, these spatial markings suggest that those creating and residing in those geographies are absolved of responsibility for the historic and ongoing occupation of Native lands, as well as the cultural and material consequences of colonization. As Herman points out in relation to Hawai'i street names, "anti-conquest is never a conscious process. Colonizers usually perceive it as paying genuine respect to the local culture, and would take offense if one were to confront them by suggesting their 'gracious acts' were in fact modes of power" (Herman 1999, 78).

The force of anti-conquest derives not only from its surface truth—that current citizens did not directly colonize the Americas—but also from its capacity to ignore corresponding cultural and material profits still accruing for all those indirect beneficiaries of anti-Native colonialism. Despite good intentions, the occupation of this continent necessarily continues to impact Native and non-Native communities unequally. My focus here on White space and White residents inhabiting Indianness helps direct our attention to anti-conquest as part of the production of space, that most fundamental medium of colonization and one of the ongoing products of Native dispossession and racialization.

I should briefly note here that some tribal communities do host names similar to those found in the White communities I have presented. In Talequah, Oklahoma, the tribal headquarters for the Cherokee Nation, the small row of seven streets using the syllabary on their signs (much like the Eastern Band in chapter 1) are also named after tribes in a way that is otherwise indistinguishable from those found in many White neighborhoods: Muskogee, Cherokee, Chickasaw, Choctaw, Keetoowah, Delaware, Shawnee. Before the Cherokee script was added, the signs offered the same physical presentation as one finds in Tualatin, Oregon, or any other White community, save the scale of the clusters. Despite this seeming parallel, they are not identical practices because of their contextual uses and meanings (however flexible). The spatial work of these names must be drawn from both the content and the relation between content, geography, culture, and history. In Talequah, the signs hold specific and not abstract meaning in relation to their understanding of indigeneity. Except for the Keetowah or Eastern Band of Cherokee, which remained in the east, all the tribes represented on Talequah street names were relocated to Indian Territory/ Oklahoma and have homelands (no longer reservations) in the area. In an intertribal space like Oklahoma, the specificity of the tribal identities listed on these signs would not be lost on the local Native residents and tribal members. Even if the signs were initially posted by non-Natives, the meaning shifts upon tribal revision and through self-legibility, and thereby comes to serve as a sign of the immediate and concrete tribal presence and survival.

In this way, the sites discussed in this chapter serve as counterexamples to those sites in the tribal communities that also use "Indianness" for street sign systems. In both cases, we can observe a material deployment of representation. Names are selected. Signs are posted. Space is made. In both tribal communities and the White-dominated neighborhoods that use such sign systems, indigeneity and Indianness signals land, identity, culture, and belonging. Both are also rooted in colonialism, and are the material outcomes of dispossession. Despite my criticisms, both are likewise the result of collective efforts to address Native dispossession, and to in some way reconcile its contradictions. Tribes seek to assert sovereignty. Non-Native communities (at least ostensibly) seek to reject the values that fueled those historic and devastating oppressions that continue to make such assertions of sovereignty so necessary. They want to, perhaps desperately, embrace Indianness. Regardless of the community use, the value of these sign systems for both Native and non-Native communities rests in helping to narrate a relationship to the lands they mark.

In my efforts at comparison, Indianness and indigeneity serve as organizing texts; but the material and discursive texts both shape and are shaped by the shared relationships, identities, and spatial understandings found in the respective communities. When used in non-Native contexts, Indianness reflects (perhaps unsurprisingly) the community identity and not the referenced (Native) communities. Indianness is mostly a projection and reflection of Whiteness. The lack of precision in the usage of Indianness, for example, provides the first point of contrast and signal to this act of projection. In non-Native communities, Indianness almost invariably signals vague and racialized notions sometimes replete with all the troubling representations of Native peoples one might expect to find in such communities. Consider again the recurrence of tomahawk in such clusters. In every case, the Indian-themed residential neighborhoods lack any concrete connection to actual Native communities.

These non-Native naming practices ultimately create another cultural sphere against which Native peoples must negotiate in order to prove and protect their tribal identities and sovereignty. Laying claim to citizenship in the Mohawk Nation or traditional Mohawk land is quite a different political and spatial assertion than laying claim to Mohawk Street. These points of clarification remind us that these street clusters are characterized by both a racially specific demographic reality and a racially and colonially informed set of constructs. Such spaces can be Indian-themed, in part, because space is not typically considered racialized except when occupied by non-White bodies.

The street names beg us to notice that Indianness and Whiteness are not in opposition, and also are not always completely different things. In many ways, each exists to give meaning to the other. Indianness and Whiteness have always been in relation to one another, especially in regard to space. Ultimately, race and culture can only make sense, as differences (or samenesses), through spatial articulations. Difference is less about distinctions per se than about prescribing where differences can exist, or not, how they will be freed or contained, how geographies will be delegated, and which meanings applied to lands will hold precedence, as well as where one can stand or travel, how one is granted belonging, what one can own, and which group(s) are dominant and which subordinate.

So, the Indian-themed sites in demographically White communities do more than simply host predominantly White residents. They reflect and (re)produce space that is explicitly informed by a Eurocentric imagination. Certainly each of the clusters offers a distinct demographic and cultural story, and the meaning of these spaces can always be recrafted with different sets of meanings and different populations. The shared use of Indianness in a neocolonial nation in which indigeneity remains strategically ignored in discussions of social justice and indigenous rights, however, largely precludes generating authoritatively alternate (anticolonial) geographic reimaginings for these sites. Instead, much like Indian mascots, they reproduce and locate Indians in abstracted ways, thus doing little to resolve the ongoing tensions and consequences of occupation, dislocation, and the "erasures" of Native peoples and spaces. Indeed, such questions rarely arise, since the power of these dominant cultural and spatial productions rests precisely on their banality. As mundane spatial markers, street names remain unquestioned as modes of hegemonic cultural production that operate at the intersection of colonialism, identity, race, and space.

Indian-themed street names do not just help us navigate roads and cities. They help us define, locate, and negotiate a neocolonial nation. They help locate Whiteness.

Indianness for Everyone

Despite the history of conquest, genocide, and assimilation aimed at destroying Native peoples and wrenching Native lands away from tribal control, Indianness regularly proves useful for negotiating White identities and a sense of belonging on this continent. Philip Deloria skillfully documents that White Americans have engaged in various forms of "playing Indian" as an attempt to resolve tensions in national, collective, and individual identities (Deloria 1998). Since the Boston Tea Party—what Deloria points to as a symbolic moment of national birth rooted in playing Indian—appropriations of Indianness have remained culturally powerful and even marketable tools for crafting White identity and frames for nationhood in the United States. The various "Indian performance options" available have helped give "meaning to Americans lost in a (post) modern freefall" and enabled them to "meet the circumstances of their times" (Deloria 1998). Men and those chasing masculinity have been especially keen on this play, which probably also partly tells us why tomahawks, rather than baskets, are so popular as Indian-themed street names.

Between the 1940s and the 1970s, desegregation, civil rights, and global military campaigns all presented a palpable challenge to White hegemony over the American cultural landscape. Although housing and homeownership opportunities were being massively expanded post–World War II, that expansion was also characterized by institutionally supported racial discrimination and exclusion. After World War II, issues of racism and structures of oppression were under scrutiny. Policy shifts starting in 1965 dramatically changed the patterns of immigration, increasing the numbers of entrants from Asia and the Americas south of the US-Mexico border. Indian-themed streets proliferated precisely as African Americans, Latinos, and Asian Americans seemed to pose cultural, political, social, and geographic "threats" to a newly established middle America. At a time when new generations of White Americans were gaining access to property and a chance to build equity for the first time, the redeployment of the Indian hardly seems surprising. Indianness offered the comfort of a connection "to the very beginnings of the mythological structure called America" and thus likely soothed White apprehensions about losing a previously secured sense of place and notion of belonging, both within and outside the nation-state (Green and Massachusetts Arts and Humanities Foundation 1975).

Between the 1950s and 1980s, western films were hugely popular, while the American Indian activism and counterculture movements repositioned Native cultures and peoples as victims of oppression. Native peoples also came to symbolize a growing environmental consciousness, and were the least demographically significant "minority" group being figured into a developing national consciousness toward multiculturalism. Indianness, it seems, could not have been more useful in negotiating the politics of racial identity and unprecedented challenges to dominant Whiteness. Indians could be symbolically incorporated with little material compromise. During these key decades, housing developers extensively applied Indian themes to their residential creations.

Deloria focuses on literal acts of "play" whereby White Americans physically or symbolically masquerade as Indians. The placement of Indian-themed street names represents a spatial variation of "play" that still presumes ownership over Indian identities but relocates that ownership along with the colonized land itself. Precisely in light of that history of appropriation, and the relative demographic absence of Native peoples in Indian-themed neighborhoods-to-be, acts of inhabiting Indianness provide the kind of spatial reconciliation (or disavowal of colonization) that marks and reaffirms contemporary White space. Moreover, inhabiting as a conceptual frame does not apply only to those who live on Indian-themed streets. Although certain individuals and families do reside in and claim those places, their experiences point to the larger cultural and symbolic space that allows developers to build and residents to dwell on Indian-themed streets in the first place. Indian-themed spaces, and the notion to create such spaces, implicate a broad cultural realm in which Indians are available for appropriation, purchase, and, in this case, literal occupation. In this way, the specific residents living in these particular neighborhoods matter less than the possibility for anyone and everyone to occupy those places. In White spaces, Indianness is turned to the task of crafting abstracted space for the supposedly abstract citizenry of the nation-state.

I want to end by emphasizing that I am not suggesting that the residents of these spaces are somehow uniquely racist or frontline neocolonialists. Rather, they are most immediately and uniquely positioned in a very real spatial sense to participate in the occupation of geographies already thoroughly constructed through an explicitly Eurocentric imagining of the American landscape. In the larger cultural realm, Indianness is ultimately available to anyone who desires access, and it is always tied to the constructions of Whiteness. Inhabiting Indianness in White communities with Indian-themed street names/signs involves laying out the simultaneous recognition and denial of Native peoples in the material and social construction of these places. In these clusters, the now-filled "emptiness" of those residential areas is produced in part by invoking easily controlled indigenous ghosts. In these communities, the spatiality of Indianness is redeployed to confer geographic belonging to the non-Native residents—in effect, to confer this belonging to all citizens of the nation without acknowledging the ongoing and corresponding sacrifice of indigeneity and Native space.

3

The Meaning of Set-tainte; or, Making and Unmaking Indigenous Geographies

Satanta, Kansas, Saturday, May 8, 2010

White beams of sunlight break through the high clouds that hover over a brisk morning. Small groups of bundled individuals hustle toward preferred viewing spots in anticipation of the day's activities. In a few short minutes, the sixty-ninth annual Satanta Day celebration will begin. Some, like me, decide to sit on the aluminum bleachers perched across the street from a semitruck flatbed. This is a prime location, since today's featured ceremony will take place directly between the bleachers and flatbed. Most of the audience members are forced to edge themselves along the curbs of Sequoyah Avenue, locally referred to as "main street." As it does every year since 1958, the Satanta Day celebration centers on the symbolic arrivals and departures of town royalty. Today, a new Chief and Princess Satanta will be honored and recognized. These town figures and the community itself are all named after the nineteenth-century Kiowa warrior, Set-tainte. This year, high school seniors Shylo Evans and Naura Harlow take center stage. They pace dramatically, moving north from the southern end of "main street." At the other end of the street, and approaching from the south, are high school juniors Kurtis Clawson and Jazmin Longoria.

The two pairs of young people are dressed in full ceremonial regalia, hand-made outfits offering their best approximation of nineteenth-century Kiowa attire. They meet at the midpoint of the street, directly in front of the Satanta Day emcee, who stands behind a humble podium and microphone. This is the moment for which everyone has gathered. All of the town is here. So are alumni, friends, and former residents stretching back several decades. The symbolic center of this day's gathering is Set-tainte.

Apache, Oklahoma, Saturday, June 12, 2010

My first visit to Oklahoma. The summer heat and humidity stifles movement. When outdoors, locals move only fast enough to make it to the next location that offers the relief of air conditioning. Just hours after I leave Oklahoma, overnight rain will create flash floods that rip through the state, flood the airport entrance, wash out highways, and trap people in trees. But today, the air is still and thick and dry.

Inside the Comanche Community Center building, which is just large enough to contain a single basketball court, the pounding of the drum reverberates through your body. The interior walls of the community center add to the visceral experience of this moment, all boldly painted to mirror the powerful blue, red, and yellow design of the Comanche Nation flag. Before being invited to sit with the host family, I soak in the scene from the top row of a small rise of bleachers. The twenty-first annual Chief Satanta Descendants powwow is under way. As with all Kiowa gatherings like this one, the gourd rattles shake almost incessantly. The dancers dip up and down, flexing at their knees, their left hands holding feather fans, their right hands rapidly flicking rattles in time with the drum. During many of the songs, a cavalry bugle commands all your senses, intermittently blaring out the iconic army call for "charge" and momentarily rendering the pulse of the drum and rattles to the background. Songs end with a noticeable call of remembrance to Red Wolf, who taught Gourd Dance to the Kiowa people (Boyd 1981, 112–115).

The songs and dancing continue into the night. People will be fed. Speeches will be given. Honorings will be made. The symbolic center of this day's gathering is Set-tainte.

Genealogies of Colonialism

As this chapter's title suggests, I am interested in the sets of meaning given to and derived from indigeneity and Indianness in the specific form of Set-tainte—a real-world historical figure who has been and continues to be simultaneously reproduced as savage, mythical ancestor, forefather, cultural hero, and tribal icon. Well over a century after his death (in 1878), this man remains a key player in the interrelated production of colonial and indigenous spaces across the southern plains. Consider Governor Sam Brownback's "Blue Ribbon Panel for Kansas

History" decision to name Set-tainte as one of the first five of the "twenty-five most notable Kansans" for the state's sesquicentennial in 2011 (Anon. 2011). The accompanying online entry hosted by the Kansas Historical Society ends with a quote from Set-tainte, implying his standing as a mythological ancestor to the settlers and all subsequent Americans: "Your people shall be our people, and peace shall be our mutual heritage" (Kansas Historical Society 2014).

Despite this framing of Set-tainte's quote, it seems clear that his statement describes a welcoming or incorporation of European Americans within a Kiowa "family" or social structure, and not a concession of a transferred identity and geography, from Kiowa to (European) American. This Kiowa-centric reframing would match the more typically bold stances from Set-tainte, who was often unmoved by the obvious and growing inequities in power on the southern plains during his lifetime.

As this quick point suggests, my goal is to outline how mundane practices of geographic imagining play a pivotal role in the processes of colonization and decolonization, with both processes being understood as ongoing, contemporary projects rather than concluded, historic ones. To accomplish this objective, I trace colonial genealogies through the annual performances in Satanta, Kansas, and the cultural assertions of a Set-tainte descendants' group based in Oklahoma. As my opening narrations indicate, the first story of this chapter centers on the Satanta Day Ceremony in rural southwestern Kansas, which annually commemorates the town name and bestows titles of chief and princess on a new pair of its residents. The second story considers the annual Set-tainte Descendants Powwow in Oklahoma and broader Kiowa efforts to remember and sanction Set-tainte's anticolonial vision for Kiowa identity and space, as well as continue their maintenance of the Set-tainte name. The juxtaposition and overlappings embedded in these stories illustrate the multiple ways that Indianness and indigeneity can be used in the production of space. The specific means by which each of these two uneasily interconnected communities remembers and embodies this man, however, is ultimately tied to whether, and in what ways, they are engaged in collective, everyday projects that are either making or unmaking Native space.

"Native space had to be unmade as much as it had to be made," Paige Raibmon astutely reminds us in describing the relationship between the transformation of the North American continent and the creation of reservations (Raibmon 2008, 58). In the case of my particular stories, I am interested in how the lands of southwestern Kansas required being unmade as Kiowa (and other tribes') lands before they could be remade as American lands, or as a re-racialized, White space. Furthermore, as I argue throughout this book, that process of unmaking Native space, or indigenous geographies, is an ongoing and necessarily incomplete process. Colonization does not stop once indigenous occupants have been removed. Nor does this process involve only large-scale policies and institutional practices (Rifkin 2013). Historically, we have seen how mundane "settler practices and Indian policy combined in a mutually sustaining dialectic to do the work of colonialism," especially during the eras of removal and reservation formation (Raibmon 2008, 58).

I would suggest that the role of those mundane practices has expanded and now largely exceeds the role of policy in serving as the central mechanism for maintaining spatial reconfigurations—the constant task of making and unmaking geography—during our contemporary era, which is removed from more direct colonial practices. While these mundane practices are less dramatic, and seemingly small-scale, they are no less effective in continuing tribal dispossession and possession-taking. They are also crucial in constantly re-narrating dispossession. To cease colonial narrations and performances of dispossession would invite questions about settlement or, worse, question settlement altogether. Thus, the residents of Kansas, or any colonized land, must continue to both unmake and then (re)make the spaces they occupy. In the case of Satanta, this is counterintuitively accomplished through a direct deployment of Indianness. Looking at a similar use of Massasoit (an eighteenth-century Wampanoag leader), Jean O'Brien and Lisa Blee describe how such appropriations recite "the origin story of predestined colonialism [that] is powerfully portable; it 'fits' anywhere and everywhere native homelands have been claimed as part of the national domain" (Blee and O'Brien 2014, 638).

The town of Satanta necessarily overlaps with the Kiowa Nation in geographic, historic, and discursive senses. Thus, my comparison of these communities allows me to explore how each uses notions of indigeneity/Indianness differently, and how so doing produces different kinds of space. The Set-tainte Descendants Powwow, like Satanta and the Satanta Ceremony, draws on spatiality, but instead mobilizes Native identities toward the production of an indigenous geography—and one that in many ways overlaps with and sometimes contradicts the geographies produced in southwestern Kansas (and many other places not explicitly addressed here). Positioned alongside one another, Satanta, Kansas, and the Set-tainte descendants offer an opportunity to more generally reflect on how the production of space (dominant, indigenous, both, or otherwise) is constituted by the intersections and overlappings of race, gender, indigeneity, Whiteness, and everyday acts of colonialism.

A Moment of Pause

Before I continue with a closer look at these two events, I want to note that there exists a rather amiable, if loose, relationship between the town of Satanta and the Set-tainte descendants. Certain residents and descendants have developed somewhat closer connections, and consider one another friends, even "family." In short, the town makes a concerted effort to directly connect their history with contemporary Kiowa peoples by inviting Set-tainte's descendants to their celebration and formal recitation of his history. Further, after his descendants recovered and reburied Set-tainte's remains in the 1960s (discussed below) the residents of Satanta pooled resources and donated an attractive headstone, which now sits in place at the burial grounds in Fort Sill, Oklahoma.

I introduce this point because it is tempting (even for me) to want to situate these two groups of people in opposition, and locate them in direct contestation with one another: a largely White community informed by settler colonial and racial logics versus a resilient and dispossessed indigenous group commendably maintaining an embattled heritage. But this would simplify the reality, even if that general outline proves relatively accurate and meaningful (as I will show). It is also certain that the core residents of Satanta are relatively powerless in the sense that they live in a small, rural, agriculture-centered community in the otherwise people-barren plains of Kansas. They are relatively poor. They exist on the margins of urban centers of power. The town offers free land to newcomers. Despite these facts, however, the town residents have benefited from and continue to benefit directly or indirectly from nodes of social power and privilege generated by race, colonization, and citizenship (among others). The free land program, for example, marking a desperate community attempt for adding community members, can also be read as a localized and modern homesteading practice once (and now once again) deployed to redistribute Native lands to newcomers.

This tension has not, it appears, limited these two communities' abilities to forge a mutually productive if at times uneasy relationship. The descendants, for example, provided the handmade shawl now worn by each year's Satanta

FIGURE 3.1 Some of Set-tainte's descendants prepare to perform a women-only scalp dance in Satanta, Kansas, 2012. Photo by author.

Princess, one identical to those worn by the descendants themselves at their pow-wow and other gatherings (see fig. 3.1). In 2012, I returned to Satanta because several of Set-tainte's descendants came to the Satanta Day ceremony to share their thoughts on Kiowa life and history. They danced and honored their most famous ancestor annually memorialized, perhaps awkwardly, by this small town. Betty Washburn, a leading elder for the descendants group, expresses genuine affection and appreciation for the opportunity this town provides: "We always look forward to visiting the Satanta, Kansas, people" she told me when I asked why they come. "We go and visit because we are Satanta's people and to honor him" (Washburn 2010).

In many ways, the honoring of Set-tainte in Oklahoma and Satanta in Kansas mirror one another. Both events involve an entire community. Each recognizes and celebrates an ethic of community service. Both frame their respective event through rhetorics of tradition. Both rely on notions of indigeneity. In both cases, a person carries the name of an important, historic figure.

Yet the distinct cultural and historic contexts of these two events are significant. The differing means of carrying the Set-tainte/Satanta name is important.

In Satanta, the chief and princess titles have changed bodies sixty times in the last sixty years (as of 2017). In that series of performances, Set-tainte and the Kiowa are ultimately mere players in the rearticulation of the town's existence via colonization. While the residents know some of the general history of their region, they maintain a superficial connection to the Kiowa community and pay little attention to contemporary Kiowa concerns. Thus, as the Satanta performers carefully pace the Indian-themed streets that re-place textual Indians back into their town, they symbolically march out and reclaim the same territory, and reperform a discursive dis-placement of the Kiowa into (just) Oklahoma. As the Satanta Day ceremony script tells us, the Kiowa simply left when they "went south toward the traditional hunting grounds" (Anon. 2012). In this way, the Satanta ceremony reflects and actively reproduces the powerful and enduring discourse of manifest destiny and the trope of the vanishing Indian, even as the Kiowa sometimes physically stand right there with them during the proceedings. The entire ceremony rather spectacularly ignores the direct relationship between their town's existence and appropriated identity, and the past and current status of indigenous peoples—the Kiowa in particular—as subjects of an ongoing settler colonialism. Regardless of the seeming contradictions of Kiowa presence and absence, the town residents who now comfortably reside on the very lands Set-tainte unsuccessfully sought to retain for the Kiowa are able to narrate and naturalize the process of colonization and to erase the racial specificities of that process of dispossession and tribal suppression; they thus affirm the town as an active neocolonial space. They do so, in part, through their conscious and ceremonial use of Indianness, although partly enabled by the willing participation and presence if not exactly the full support of Set-tainte's descendants and other Kiowas.

While I argue that Set-tainte and the two events centered around this man play a role in the simultaneous making and unmaking of non-Native and Native spaces, it should be clear that none of his descendants have ever suggested to me, or said in any public moment that I am aware of, that the Kiowa (or these descendants specifically) wish to reclaim the plains of southwestern Kansas. The impracticality of such a potential scheme renders this kind of proclamation unnecessarily burdensome and unrealistic (at least for now), while more immediate, pressing tribal needs clearly takes precedence over land claims already negotiated, rejected, or "paid for" by the US government through the Indian Claims Commission (US Indian Claims Commission 1979). Nevertheless, I hold steady in my position that Set-tainte's descendants and the Kiowa more generally would agree with Set-tainte if pressed on the question of whether those lands south of the Arkansas River in Kansas fall under the purview of their hereditary territories, and thus are still envisioned as Kiowa space.

The tension of these spatial truths press against one another in a way that parallels the stories of Set-tainte's final demise. Mainstream historical documents overwhelmingly and continually report that he jumped to his death from a prison hospital window, never resigned to an inevitable fate of lifelong imprisonment. Kiowa narratives insist Set-tainte would never have committed suicide, and thus imply in culturally appropriate understated fashion that his death came at the hands of American officials. For my purposes here, the "facts" of Set-tainte's suicide/murder matter less than the positions and beliefs constructed on either side. No Kiowa are seeking retribution or even public proclamation of their ancestor's killing, but their understanding remains firm. The stories are recounted. Set-tainte's stance is protected, and perhaps extended, in this way. In a similar way, Kiowa geographies that overlap with Satanta, Kansas, remain in tension. Thus, we must concede that Kansas (along with parts of Nebraska and Texas, as well as their current home in Oklahoma) remain part of an imagined, traditional Kiowa geography. This also means that Satanta and many other places necessarily exist as overlapping indigenous and colonial geographies. While remade, they have not been fully unmade as Native space.

Set-tainte and the Founding of Satanta, Kansas

Set-tainte, the individual, was a nineteenth-century Kiowa warrior and political leader who fought against White oversettlement and the concurrent destruction of Kiowa and indigenous ways of life on the southern plains. Standard historical narratives label Set-tainte one of the "militant" Kiowa leaders who, despite his involvement and presence at numerous treaty negotiations, actively resisted concessions offered by other "peaceful" Kiowa groups and leaders.

As a prominent tribal figure who served as a principal chief in the 1860s and 1870s, he was notably present for the Medicine Lodge Treaty discussions in 1867. During the treaty proceedings, St. Louis Daily Republican correspondent William Fayel wrote Set-tainte's "name is a terror to the whites of the frontier. . . . The Kansas people attribute many of the depredations committed to this Indian, and think he richly deserves a hempen neck tie" (quoted in Robinson 1998, 63).

From this unlikely candidate came a town name in 1912, when the Atchison, Topeka and Santa Fe Railway company (AT&SF) manufactured Satanta along its westward line out of Dodge City. Although Set-tainte's name no longer registers in contemporary public memory, it still resonated deeply and negatively with European Americans in the early twentieth century, many of whom, like Fayel, considered him a "snake" and a "scoundrel." In one neatly coordinated gesture, the railroad company named the town after an undisputedly anticolonial indigenous person, and thematized its streets using Indian tribal designations such as Nez Perce and Apache (see fig. 3.2). This assessment remained at the time of Satanta's founding. When the *Haskell County Republican* announced the naming of Satanta, the writers reminded their readers that the town's namesake was "an Indian chief that led bands of marauding Indians into Western Kansas," conveniently neglecting the historical fact that the Kiowa counted those portions of Kansas as part of their traditional territories.

In addition to stating that Set-tainte was a "distinguished orator" and key signatory to the Medicine Lodge Treaty of 1867, the writers recalled that "on a raiding trip in Texas in 1871—a band of Kiowas under Satanta attacked a wagon train and killed seven white men" before they were arrested, tried, and sentenced to life in prison. While not articulating any explicit resistance to the town's name, the writers made sure to be clear on the dominant perceptions of Set-tainte's racialized legacy in Kansas, asserting that "early settlers in the southwest will recall old chief Satanta as one of the scourges of the plains" (Anon. 1912).

Looking back, the use of a Native-inspired town name is not historically unique. The AT&SF promoters alone produced no less than forty-two towns with so-called Indian names—plus one in Texas that was intended to be named Satanta but instead was mistakenly transcribed as Santa Anna (Marshall 1945, 357–358). The railroad company actively capitalized on and marketed interest in the art and lives of Native peoples, perhaps most famously deployed by railroad entrepreneur Fred Harvey (Dilworth 1996). His famous "Indian Detours" generated great profit for many southwestern towns and entrepreneurs and helped spark a national art collecting interest in Indian crafts during the late nineteenth and early twentieth centuries. While the creation of Indian town names and the general use of Indianness to market all sort of products was already common by the early 1900s, the comprehensive spatial ordering (via street names) of Satanta through reference to the very people indigenous to those lands was unique. In fact, none of the other forty-one towns to which the AT&SF gave

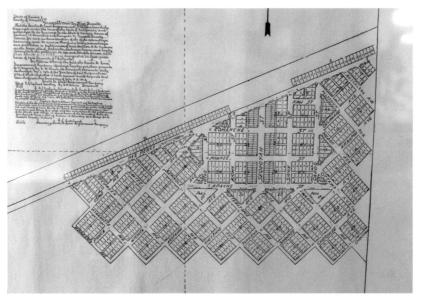

FIGURE 3.2 Detail of Satanta town plot filed by the Santa Fe Land Improvement Company, 1912. Courtesy Dudley Township Library.

Indian-inspired names also housed or has since come to house any collection of Indian-themed street names. Satanta alone used this theme, and it did so fully and extensively.

In Satanta the theme extended to the names of local shops and other commercial enterprises, such as the "Big Chief Garage, the Pocahontas Theater, the Hotel Modoc, and . . . Wigwam Meat Market" (Worster 1979, 167). Southern Plains historian Donald Worster notes that the Atchison, Topeka and Santa Fe Railway promoters crafted thematic street names (like suburb divisions today) for both nearby Sublette and Satanta. Sublette packaged its roads with the names of "Western frontier heroes, many of whom had never seen the area— including [Henry] Inman, Carson, Cody, Lalande, and Choteau" (Worster 1979, 167). Satanta, he chides, "was getting its name from a Kiowa chief who had been driven to alcoholism by defeat and had died in a federal penitentiary. The promoters applied the same thoughtless gimmickry [as Sublette] to its checkerboard of wide dusty avenues: Sequoyah was the main commercial street, after which came Apache, Comanche, Kiowa, and so on" (Worster 1979, 167).

Worster critiques what he considers "replication of clichés" producing meaningless differentiation. "However colorful as merchandizing strategies," he tells
us, "names such as these, borrowed wholesale from an alien past, could not dis-
guise the true identities of these towns. In reality, both Sublette and Satanta were
readymade versions of Main Street" (Worster 1979, 167). Apparently, early residents shared Worster's sentiment, if for differing reasons. Most local historians
indicate that many citizens were upset by the AT&SF's choice of "Satanta" and
claim that their elders did not like the idea of a White Kansas farm town forged
out of the Great American Desert being named after a Native warrior, especially
one considered as notorious as Set-tainte. These sources insist that the negative
response was lively enough to convince the powerful railroad company to forever afterward relinquish its town-naming rights to locals. Today, we might consider whether this narrative and perhaps the subsequent ceremony both act as
"evidence" to shield against current implications with racism and unwholesome
occupation. In any case, by the mid-twentieth century, the use of Indianness in
Satanta expanded beyond the mundane spatial markers and commercial ventures to include the spectacle of an annual "crowning" ceremony complete with
town-wide festivities and a parade.

Satanta Day Celebration 2010 and 2012

Local resident Jackie Hall narrates Satanta Day and serves as officiant for the
chief and princess ceremony, a position she has held for twenty years. In the
introductory remarks she reminds her audience that,

> it is a great honor indeed for a young man and a young woman to be chosen to
> represent this community in such prestigious positions, but more important than
> this, these young people have been chosen as symbols to perpetuate the memory
> of the first Satanta, a great chief and warrior of the Kiowa Nation, after whom this
> community is named. (Anon. 2012)

This cultural and spatial script, if not this precise narration, has been performed in Satanta for more than half a century, and usually unfolds to the audible satisfaction of the crowd. The emcee reiterates for the audience the significance of this gathering. On this day, Chief Satanta and Princess Satanta's names
will be carried on, although they will change bodies. Together, the new chief
and princess will hold court at a variety of local functions, representing their

town across the region and developing leadership skills. They are central figures and will play their unique part in this collective performance of consolidating the town's attachment and affiliation with its name, reifying its mythical relationship to the town's namesake, Set-tainte. Following Blee and O'Brien again, we see that Set-tainte is also being rescripted as a "protopatriot," ready for solidifying an authentic national past and securing its future (Blee and O'Brien 2014, 637).

Carefully following Hall's prompts, the "reigning" chief and princess hand over their symbols of rulership. In 2010, Shylo Evans holds the peace pipe up to the sky, "smokes" from it, and then transfers it to Curtis Klawson (see fig. 3.3). Then, each removes his headdress. Evans hands over his more elaborate "badge of authority" to Klawson, in exchange for a plainer white-feathered band. This scene is replayed each year with new sets of students. For each iteration, we in the audience are explicitly asked to view this ceremony across culture, space, and time. This event is more than just masquerade. These two individuals, the chief and princess, and all of us in the audience are to be collectively transported into the "appropriately" sexed and gendered bodies of Kiowa peoples standing here in times past, in this very place, on "former" Kiowa lands. Hall guides us to

> let the time roll back to an era when the buffalo roamed the plains, a time when the great cities were only a wilderness. Listen to the throbbing of the great medicine drum. Forget this present day life and envision in your minds the image of the youthful Satanta as he approached from the North to be proclaimed Chief of the Kiowa Nation. (Anon. 2012)

In our temporal travel and spatial imagining, and as we are "hailed" into inhabiting Set-tainte's and other Kiowa bodies, we are asked to construct a patriarchal, masculine, and White racialized space. Through the ceremonial script, we are informed that "the original Chief Satanta" possessed "a manly boldness and directness," a description found in local journalist Gladys Davis's 1930 article, although she indicates this phrasing is derived from government reports on Set-tainte (Davis 1930). His humor and eloquence, we learn, is said to have gained favor among army officers and commissioners "in spite of *his* hostility to the white man" (Anon. 2012; emphasis added). In a decidedly less ceremonial and highly heteronormative and gendered conclusion, outgoing 2010 Princess Naura Harlow hands over her shawl, necklace, and ring to Jazmin Longoria. As with each year, she is given nothing in return. For this portion of the ceremony,

FIGURE 3.3 The incoming Chief and Princess Satanta receive gifts from their outgoing counterparts. Satanta, Kansas, 2010. Photo by author.

the emcee indicates that we are finally given liberty to use our imagination, as we "recall the exchange that *might have* taken place between the bride of Satanta and those that would wish her well and good luck in assuming *her position beside such a great man*" (Anon. 2012; emphasis added). Here, it is clear, the princess is relegated to recognition merely by being in proximity to the Chief's "manly boldness and directness."

Such historically rooted rearticulations emphasize Indianness as masculine precisely the way they did during the era of westward expansion. Precisely the way wilderness is constructed as space of masculine endeavor, in need of submission to human hands (Evans 2002). This implied contest over masculinities has long proven central to narrations of hard-earned settlement and conquest of both Indians and the land, with Indian men ultimately proving less manly than White men. Finally, no mention is made of which of Set-tainte's wives we should imagine during this portion of the ceremony. He is understood to have had up to four at the same time. During the ceremony, we are not encouraged to dwell on exactly what kind of ceremony this might represent, if we did acknowledge and account for the traditional Kiowas' polygamous practices.

Despite tensions in script, the narrative establishes that the audience now occupies the reimagined bodies of the Kiowa, and thus, standing in the same landscape as the Kiowa once did, it seems appropriate that "we" now occupy their lands as well. At this delicate moment of constructed timelessness, which is marked by a fabricated racial and cultural fusion, we find colonial spatial constructions in action. The temporal and temporary remaking of a Kiowa geography is (again) "destined" to be unmade upon the ceremony's conclusion. Just as an Indian identity transfers bodies during this performative moment, so too does the land transfer ownership hands. These two transfers are not, however, equally sustained. At the end of the ceremony, the appropriated and imagined Indian identity is officially released by everyone except the chief and princess. The land remains claimed. Settled. Occupied. Inhabited.

It is still not clear how or why the town started proclaiming a Satanta Chief and Satanta Princess in 1958 (seventeen years into the Satanta Day celebration, which began in 1941), but the decision was bolstered by the visit of a Kiowa man in 1959. In that year, Robert Goombi came to Satanta with a troop of dancers, advertising for the annual American Indian Exposition held each summer in Anadarko, Oklahoma, since 1932. Town historians proudly report that Goombi not only attended the Satanta Day festivities, but that he indeed conducted a "ceremony" that "officially made all Satanta citizens honorary residents of the Kiowa" Nation (Anon. 2013). Although Chief Goombi's role clearly lent the Satanta festivities an air of legitimacy in terms of its appropriated Indianness, his "official" honoring had not stopped the residents from crowning their first "Satanta Chief" and "Medicine Man" the year before his visit; a practice that continues to this day and no doubt would have continued absent Goombi's support. Indeed, we see this mythical transference already described by Gladys Davis in 1930, when she suggests that the town's "Indian spirit still prevailed" after the removal of Native people. She ends her historical piece saying, "Welcome to our Indian village where teepees have blossomed into modern bungalows and white 'savages' await with big hearts and open hands to greet you" (Davis 1930).

When I asked about Robert Goombi's role in naming the citizens of Satanta honorary members of the Kiowa Nation—a celebrated fact that is embraced as a marker of the town's long-standing and approved manifestation of the Kiowa (and Set-tainte, specifically)—one important elder and Set-tainte descendant explained that Goombi did not "have this right" and that he actually had "no blood nor relationship ties to Set-tainte" (Washburn 2010). Goombi's and

Satanta's respective economic and cultural motivations seemed to collude during that much-vaunted visit.

The Satanta citizenry's striking and elaborate claim to Indianness must be viewed as extraordinary, but we should note that their claims had at least once before extended beyond just spectacle and masquerade—and right into the grave. Mirroring current repatriation politics and genomic research tensions, Satanta's legacy of appropriation—their use of Indianness—convinced city leadership to assert claims over not just Native history, identities, and land, but also actual Native bodies.

After his death in 1878, Set-tainte was buried in a Huntsville, Texas, prison cemetery, where his body stayed for the next eighty-five years. His remains were eventually reclaimed by the Kiowa Nation in 1963. Set-tainte's grandson, James Auchiah, led a successful effort to recover and reinter his remains in Oklahoma, despite a few tense moments, including Texas representative (Weatherford, Texas) James Cotten's asinine proclamation that "Satanta led numerous raids on the white settlements in Texas, *including some in my district*" (Stanley 1698, 351–352; Robinson 1998, 197; emphasis added). Cotten went on with his anachronistic conflations that prefigured contemporary American political boundaries, saying, "Satanta was a murdering, thieving Indian. He killed a lot of white people in my area."

As Set-tainte's descendants were patiently working to have their ancestor's remains returned to the Kiowa people for proper reburial, Russell Winter, the president of the Satanta Chamber of Commerce wired a "claim for the body" and argued that the town should "get the bones of Satanta" (Stanley 1698, 352–353; Robinson 1998, 197). The basis of Russell's claim rested confidently on the name of the town, and the fact that many of Set-tainte's "nefarious" acts of resistance took place on "the high plains of western Kansas" where the town of Satanta now sits. Winter hoped to receive the remains in time for the twenty-second annual Satanta Day celebration, during which "re-burial could be appropriately carried out" (Stanley 1698, 353). Clearly, the residents of Satanta, Kansas, took their Indianness seriously. Kiowa geography was openly recalled but only just enough to be deployed as the basis for establishing Satanta's lineal and spatial claim to Set-tainte. When I asked Russell Winter about this in 2012 (at age ninety), he apparently no longer recalled that event. Perhaps his failure was not worth remembering. The Texas State Legislature and its Department of Corrections ultimately rejected the town's claim. In this framing, however, we see how Indianness is about possession, and it is about recasting settler occupation as either metaphoric or literal descent (Ben-Zvi 2007). Ironically, then, Indianness in this framing is about Whiteness and White space, and an identity that is not linked to the living, tribally enrolled Kiowa peoples now firmly displaced to Oklahoma.

The Satanta claim to Set-tainte's body can make sense only because it is located on the site of his, and the Kiowas, dispossession. This seems the very definition of colonial ambivalence. On the one hand, this historic figure—and in some ways the Kiowa at large—are incorporated into the narrative of the town and nation. At best, this reflects a well-intentioned but flawed effort at inclusion and multiculturalism, dependent on a racialization of Native peoples that ushers them into the US nation-state as another citizen-subject. On the other hand, this symbolic incorporation reasserts dispossession. The process of Indianization described by Jodi Byrd simultaneously dismisses indigeneity and legitimate Native claims to land (Byrd 2011, xxiii). So, at worst, the adoption of Set-tainte as a historic-mythic figure by Satanta reflects an uncritical appropriation and a symbolic form of racializing and cultural violence. This capacity for appropriation, and its specific deployment for producing racialized, colonial narratives, is the essence of imperial notions of propriety and the logic of dispossession. As Reardon and Tallbear argue, these are precisely the kinds of "enfoldments" that still allow scientists today to claim Native peoples' DNA as their "rightful inheritance" (Reardon and Tallbear 2012, S234). Just as the land has changed hands, the idea that Native identities and bodies can unproblematically change hands is an ongoing sentiment found in all repatriation battles when Native remains are claimed as the property of a universal humankind. In terms of a spatial enfoldment, the claim to Set-tainte serves to again unmake Kiowa geography and to implicitly reaffirm and naturalize its replacement with a geography violently crafted through a gendered and racialized settler colonialism.

Set-tainte's Spatiality

Today, Kendall Washburn, who is the great-great-great-grandson of Set-tainte, carries the name Set-tainte. He is celebrated along with the historic man. This name/title was passed down to him in 1992 through the male heirs in his family. He earned the honor of carrying the name via his lineage and in recognition of his readiness to offer the proper respect and shoulder the necessary community responsibilities expected from the one who carries such an honor. As anthropologist Michael P. Jordan points out, "the privileged place that narratives of prominent nineteenth century warriors occupy in the historical consciousness . . . of the contemporary Kiowa community reflects long-standing Kiowa practices of commemorating male martial achievement (Jordan 2011, 46). As Kendall Washburn holds the title of Set-tainte, he also represents a Kiowa geography that is not merely a response to European American productions of (colonial) space. Although not a home of their choosing, over the last century and a half, the Kiowa have successfully balanced the celebration of their struggles to claim and retain their lands and ways of life with the practical need to make their Oklahoma site of exile/restriction into home.

While Washburn plays a current role in this cultural articulation, figures like Set-tainte have allowed the community to productively hold such cultural and spatial tension for generations. Set-tainte regularly spoke against and acted in direct response to growing European American efforts to unmake Native space in the mid- to late nineteenth century. During the negotiations at Medicine Lodge, Set-tainte drew on emotion-laden language that spoke into existence a comprehensive and specific Kiowa geography. His juxtaposition of settler and indigenous spaces included opposition to the schools and churches the government intended to provide as part of an assimilation campaign being built into treaty negotiations, and implied a relationship between the Kiowa and the land that went beyond mere resource access (bison, timber, etc.). His articulation of Kiowa space argued that one could not simply replace or abstractly substitute an alternate (reservation) territory in place of Kiowa lands. He told the officials,

> All of the land south of the Arkansas River belongs to the Kiowas and Comanches, and I don't want to give away any of it. I love the land and the buffalo and will not part with it. . . . I don't want any of these medicine Lodges [i.e., schools and churches] built in this country. I want the papooses brought up exactly as I am. . . . I have heard that you [Americans, US military] intend to set apart a reservation near the mountains. I don't want to settle; I love to roam over the prairie; I feel free and happy; but when we settle down we get pale and die. . . . A long time ago this land belonged to our fathers; but when I go up to the [Arkansas] river, I see camps of soldiers on its banks. These soldiers cut down my timber, they kill my buffalo; and when I see that my heart feels like bursting; I feel sorry. (Quoted in Robinson 1998, 70–71)

It is in the context of such speeches that I would suggest we can doubly read Set-tainte's dubbing as "Orator of the Plains." At one level, he obviously earned this distinction because of his eloquence and outspokenness during important council meetings and negotiations. At the same time, this title can also indicate his extraordinary efforts to articulate a Kiowa geography and to literally speak of and for the plains. As a leader who clearly understood that the survival and even the Kiowa's very identity were inextricably linked to the land, his orations were aimed at an assertion of an indigenous spatiality that accounted for more than just human desires. He based his resistance on an indigenous geography that acknowledged the presence and precedence of an interconnected landscape filled with a vast family of beings.

Following the Medicine Lodge Treaty, the Native peoples of the southern plains were compelled to cede certain lands to the United States, although they were assured of continued hunting and resource rights. In numerous councils and treaty gatherings, Set-tainte provided unambiguous testimony to the intentional geographic transformations being generated by Whites that threatened to render the southern plains indecipherable to traditional Kiowa spatial imaginaries. One lead official, Missouri senator John Henderson, conceded that the plains tribes could continue their hunting, although he also made it clear that the "buffalo will not last forever," and thus promised that the establishment of the reservation would provide the tribe's future security. He told tribal representatives,

> They [bison] are now becoming few, and you must know it. When that day comes, the Indian must change the road his father trod, or he must suffer, and probably die. . . . We do not ask you to cease hunting the buffalo. You may roam over the broad plains south of the Arkansas River, and hunt the buffalo as you have done in years past, but you must have a place you can call your own. (Robinson 1998, 72)

General Winfield Scott Hancock conveyed the same sentiment to tribal leaders during a meeting at Fort Larned prior to the Medicine Lodge gathering, saying "You know very well that in a few more years the game will go away. What will you do then? . . . You should cultivate the friendship of the white man now, in order that he may be your friend when this time comes" (Robinson 1998, 57). Putting aside the irony of "friendship" and concerns over tribes having a "place you can call your own," these proclamations—really explicit warnings and thinly veiled threats—highlight the active process of unmaking Native space and the rhetorical fluidity of narrating White innocence in that process.

The precise European American intentions and activities directed toward eliminating the bison herds and cutting timber actually proved a coordinated guarantee against long-term Native access to hunting and resources. Starting in 1874, White travellers and settlers flooded into the bison grazing territories, destroyed the herds, and disrupted the southern plains tribes' ability to independently sustain themselves, while the government offered tribal lands to railroad companies to facilitate settlement and commerce. By the 1870s and 1880s, respectively, the southern and northern herds were reduced to a few hundred animals, compared with an original estimated population of at least thirty million. Private citizens and settlers repeatedly ignored or actively broke government policies and treaty agreements with tribal Nations, while the government in turn ignored but ultimately benefited from the citizens' violations.

Patrick Wolfe reminds us that nonmilitary and non-government-sanctioned citizens served a crucial role within the diverse and efficiently haphazard set of mechanisms for dispossession, all of which followed the colonizing "logic of elimination." The physical assertion of an a priori dominion over and above any "mere" right of occupancy subjected Native peoples less to "state instrumentality but [more to] irregular, greed-crazed invaders who had no intention of allowing the formalities of federal law to impede their access to the riches available in, under, and on Indian soil" (Wolfe 2006, 391). When Native peoples like the Kiowa had little choice but to respond with force in order to protect their way of life, the government and media would happily identify the resistance as a betrayal of treaty agreements and the general state of "peace" and "order" supposedly brought by European American settlement. Military action inevitably followed, with the army rounding up insurgent tribal leaders and "negotiating" further land concessions. Whites then narrated the recovery of a peaceful geography temporarily disrupted by recalcitrant savages.

In many ways, the bison extinction campaigns rendered treaty concessions redundant. All of the government and settler parties were clear that, without bison, the tribes could not maintain their ways of life. Thus, the bison were targeted and railroad plans were extended. As Charles Robinson reminds us, "President Andrew Johnson saw the fighting on the plains as a threat to national development, particularly with construction of the trans-continental railroad underway" (Robinson 1998, 59). In 1867 (same year as the Medicine Lodge Treaty), a new congressionally formed treaty commission charged with the protection of railroad development across the plains intended to carve out a safe corridor by forcing all Native peoples into areas either north of Nebraska or south of Kansas. Without traditional subsistence practices, Native space could be effectively unmade, and remade as non-Native or settler space. Native resistance could then be recast as tribal inability to adjust to the "properly" human-ordered space of a colonized landscape.

Both everyday settler practices and official Indian policy reflected this shared colonial vision of the land as inevitably being occupied by European Americans and "put to use," rather than being ceded to bison herds and wandering Indians. The collective mechanisms of dispossession reflected the racially coded realities of colonization, and therefore must be read as an active example of making colonial space. Thus, despite Senator Henderson's and General Hancock's articulation of the destruction of the bison and the loss of timber lands as a passive occurrence (i.e., they "will not last"), treaties such as the one at Medicine Lodge marked the racialized effort directed toward unmaking Native space and the de facto collaboration between government policies and settler activities in manifesting a White, colonized geography.

Set-tainte and the Kiowa were entangled in an explicit moment of transformation during the mid-nineteenth century, in which the Kiowa resisted participating in the unmaking of their own geographies. Set-tainte opposed the extension of the railroad through the southern plains, refusing to concede the land. In most of the southern plains treaty negotiations during this era, the Arkansas River (which runs from west to east across the lower third of Kansas, approximately forty miles north of Satanta) served as an important territorial marker for Kiowa (and southern plains tribes') lands. Historian Charles Robinson points out that, despite the recognition of a Kiowa geography implied by this marker, two of the key Kansas representatives—Governor Samuel Crawford and Senator Edmund Ross (a director for the Atchison, Topeka and Santa Fe Railway)—did not intend to allow Set-tainte to continue the itinerant Kiowa lifestyle in "*their* state" (Robinson 1998, 71; emphasis original).

These representatives clearly signaled that European Americans were confident in the unmaking of Native space, although the need to proclaim this position also betrays lingering uneasiness about the security of such a claim. Set-tainte clearly understood the racialized nature of the American practices and policies of dispossession. In a Fort Larned meeting prior to the Medicine Lodge gathering, for example, he stated outright that "The Cheyennes, Kiowas, and Comanches are poor. They are all of the same color. They are all red men. This country here is old, and it all belongs to them" (quoted in Stanley 1968, 355–356). These simple recognitions and refusals of the colonial logics of manifest destiny and White racial superiority reflected a tense overlapping of space. The refusals of Set-tainte and others forced American officials to utilize alternating means of negotiation, conquest, removal, or assimilation to remove indigenous presence and erase Native space.

The Descendants Powwow and Spatial Production

Although the contemporary Set-tainte descendants powwow might be viewed solely as an event of cultural maintenance or revitalization, I argue that this Kiowa-centered affair also actively addresses the issue of colonization and the question of land that is absent in the activities in Satanta. For the Kiowa, as I suggest in the preceding brief history, Set-tainte is respected precisely because of his resistance to the American colonial project, and his resistance was undeniably directed at forces that explicitly sought to unmake Native space in order to make settler colonial space. Indeed, even the inclusion of the US Army bugle call in gourd dance songs honoring Set-tainte (noted in the opening narrative) counterintuitively represents the pride that Kiowa hold for their concerted resistance efforts and this figure's leadership in trying to maintain their traditional culture and lands. John Bierhorst notes that bugling songs acquired their "present form in the middle of the nineteenth century when Gourd Clan members in their skirmishes with the U.S. Cavalry vied with each other for the honor of capturing a live bugler. If the bugler refused to play at the Clan victory dance he would be killed. If he played well he might be released unharmed" (Bierhorst 1979, 8).

During one battle, Set-tainte famously killed an army bugler and stole his instrument. He subsequently taught himself to play the bugle directives and gained a level of immortality by causing confusion among the American soldiers as he ordered them to "charge" or "retreat" at his own discretion and to the advantage of the Kiowa warriors and their indigenous allies. This kind of tactical foresight and courage highlights Set-tainte's commitment to his people and their way of life, as well as his determined resistance to containment on reservation lands. His very notoriety rests on the skill of grasping control over geography, precisely in those spaces where the American military sought to remake the landscape. His actions also represent larger and ongoing Kiowa land claims, which were only finally "resolved" in 1978, when the Indian Claims Commission awarded the Kiowas, Comanches, and Kiowa-Apaches more than $43 million for lands ceded in Colorado, New Mexico, Texas, Oklahoma, and Kansas (US Indian Claims Commission 1979, 54).

The Kiowa ceremonies honoring Set-tainte today serve as reminders forged through the historic experiences of dislocation, struggle, violence, and loss. They are not merely nostalgic performances, nor playful acts of appropriation. According to John Bierhorst's notes in the Smithsonian recordings, *A Cry from the Earth: Music of the North*, gourd dance songs were "old-style Indian songs" that "reinforce Indian pride and helps to strengthen Indian claims to separateness and self-determination" (Bierhorst 1979). While Kendall Washburn carries the specific honor and responsibility of the Set-tainte name, all of his people hold Set-tainte's story, and emerge from its maintenance. This kind of historical consciousness has long reflected a Kiowa understanding of tradition (Jordan 2011). Drawing on his extensive experience with the Kiowa and Set-tainte descendants in particular, historian Michael P. Jordan tells us, "In [Set-tainte's] opposition to Anglo-American encroachment and his efforts to preserve the Kiowa people's autonomy and way of life, members of the Chief Satanta (White Bear) Descendants see a model for contemporary Kiowa people" (Jordan 2011, 118). This model, and the need for continued remembrance of noteworthy ancestors, has been an ongoing and predictable cultural practice among the Kiowa. At the end of the Medicine Lodge Treaty, for example, Satank, the leading elder representation for the Kiowa, left the American contingent with a final thought that foreshadowed Set-tainte's and many other Kiowa figures' ceremonial remembrances.

> I shall have soon to go the way of my father, but those who come after me will remember this day. It is now treasured up by the old, and will be carried by them to the grave, and then handed down to be kept as a sacred tradition by their children and their children's children. (Robinson 1998, 76)

In this light, by collectively holding Set-tainte's story, the Kiowa maintain a firm grasp on their history of struggle over the making and unmaking of space, and thus at least render Native and non-Native spaces as overlapping and in tension with one another.

An indigenous Kiowa geography persists in Oklahoma, but also in Kansas and other locations where the Kiowa no longer collectively reside or maintain a recognized community presence. This persistence denies its complete replacement. Contemporary Native space continues to defy the spatial absoluteness, certainty, and singularity that colonization intends to generate. Native space maintains layered geographies, and provides for coexisting partialities. This is one of the most important legacies created by Set-tainte, and proudly sustained by his descendants and those many Kiowa who continue to honor him.

As Set-tainte biographer Charles Robinson also rightly notes, Set-tainte's story "is a living story, carried on by living people" (Robinson 1998, 198). Indeed, one elder told me, "the Kiowa exist today only because of what Set-tainte did." Not *despite* his resistance as one of the "militant" leaders, but only *because of* his resistance. Because of Set-tainte's resistance, they still exist as a people.

Kiowa Kansas

I decided to conduct research in Satanta, Kansas, after I learned about their annual Satanta Day celebration. I wanted to gather a better sense of how this community understood its use of Indianness. I was quite fortunate that this trip organically introduced me to the Set-tainte descendants as well, as it opened the prospect for a more directly comparative analysis. Satanta has proven unique in that it actually disrupts my analysis of how the dominant spatial production of Indianness generally follows a mundane and therefore more hegemonic path, and normally sinks articulations of Indianness into an abstracted, routine, and apparently "apolitical" phenomenon—like street names and the corresponding material culture, the physical street signs themselves. Satanta does this, but it also works against this.

Communities and individuals across the nation (and the globe) habitually "play Indian" (Deloria 1998; Green and Massachusetts Arts and Humanities Foundation 1975). Chapter 2 likewise documents the prolific and massive use of Indian-themed street name clusters. These two related practices—playing Indian and inhabiting Indianness—rarely come together. Satanta is the only town that I have found that thoroughly and explicitly combines these two modes of deploying Indianness. Satanta, then, offers a rich opportunity to consider the material, visual, and performative aspects of mundane objects like town and street names and physical street signs, against spectacles like parades, costumes, and ceremonies, all of which perform multiply racialized identities and build community.

Although in many ways Satanta provides an outlier or anomaly to my research findings about Indian-themed White spaces, that community's coordination of these two deployments of Indianness also offers an instructive case study. On the one hand, the Satanta example recalls the dictum of the exception that helps prove the rule. At the same time, it also serves to make explicit the broader linkages among the colonial project of American nation-building, playing Indian, and the colonial and racialized production of space. While opportunities for collaboration and even friendship are generated, the Set-tainte's descendants do not forget the relationships of power forged through and by colonization. The Satanta ceremony, on the other hand, implicitly celebrates conquest, and thus maintains a colonial geography that tends toward Native elimination.

According to Jordan, the Set-tainte descendants slogan, "Linking the past with the present to determine the future," means the organization is focused on how the past, and how remembering their ancestors, can help them proceed and prosper in the future, as Kiowa people (Jordan 2011, 108). When Jordan asked Kendall Washburn, "how he would feel if another community member claimed the [Set-tainte] name," Washburn explained,

> I probably wouldn't take too kind to that. And you know, what can I do? What can I do? From the Indian point of view, there's no legal action or nothing like that. It's just recog . . . It would just be recognized amongst the Kiowas as bad show, you know on their part. But yeah, I probably wouldn't, I probably wouldn't take too much liking to that. (Jordan 2011, 277)

Following Washburn's position, Jordan rightly points out that traditional Kiowa law does not provide "any formal mechanism for resolving the dispute" even within the Kiowa community, and thus "individuals place their faith in public opinion" (Jordan 2011, 277). If there exists no mechanism for resolving this kind of conflict within Kiowa society, there is little expectation that a similar breach of protocol could be resolved outside of a tribal community context. Thus, when I asked members of the Set-tainte descendants group their response to the Satanta performance and claiming of the Satanta name, they indicated that their attendance at the festivities served their own purposes (honoring their ancestor) and, in true tribal fashion, they metaphorically "rolled their eyes" at the annual spectacle of playing Indian.

I saw one other clear example of this indirect approach in 2012. To kick off that year's Satanta ceremony and parade, the Set-tainte descendants performed a Scalp Dance. According to the script they gave the Satanta Day emcee, this dance is a rare one traditionally "performed by women only and only when a battle was successful and all the warriors returned safely" (Washburn n.d.). In Betty Washburn's introductory comments to the town residents before dancing, she graciously thanked her hosts, saying, "To the citizens of Satanta, Kansas, we say,

'Thank you for your kindness and respect. We are happy to meet with you again'"
(Washburn n.d.). In closing, however, she reminded the town of Set-tainte's line
of descent and their indigenous claim to Kiowa identity and history, as well as
Set-tainte's name. She continued, "Thank you for the honor of presenting a part
of *our history* to you," before unswervingly signing off as "Chief Satanta's People"
(Washburn 2010; emphasis added).

Despite Satanta's illustration that colonialism can deploy Indianness toward
the maintenance of White and settler colonial geographies, this town offers
promising opportunities for collaboration and reconciliation. In the end, the
descendants' mere presence at the Satanta Day celebration provides a sufficient
and culturally appropriate response, even if it is often read as a tacit approval
or authentication of the town's ceremony. As Kendall Washburn implies, self-
correction offers the only mechanism of resolution available. Thus, it remains
the responsibility of the Satanta residents to resolve their relationship with Set-
tainte and the dispossessed Kiowa peoples. While this may seem a daunting or
unlikely outcome, such reconciliations have occurred elsewhere. In Raibmon's
"Unmaking Native Space," she opens telling the story of a meeting between the
Twombly/Gray family and the people of Opitsaht village on Vancouver Island
(Raibmon 2008). She astutely offers this family's ceremonial apology to the
Tla-o-qui-aht as a model for beginning the process of accepting responsibilities of a shared colonial inheritance (Raibmon 2008, 78–80). I would suggest
that despite the appropriative nature of Satanta's ceremonial performance and
deployment of Set-tainte as a mythical ancestor figure, which largely ignores
the colonial and racialized implications of their spatial productions, this town's
continual and explicit engagement with Set-tainte and his descendants offers a
recurrent window of opportunity that may purposefully be transformed into the
kind of meaningful reconciliation that occurred on Vancouver Island.

A number of individuals, including most of those who have developed friend-
ships with the Set-tainte's descendants, have the interest and capacity to begin
this kind of exchange, although they may not immediately know how to proceed
properly. It seems clear, though, that many of the key persons involved, myself
now included, have the responsibility of taking the next step and executing this
important task at hand.

FIGURE 4.1 *Occupied Territory* (2013), Chris Pappan. Courtesy of the artist.

FIGURE 4.2 *Free Homes in Kansas* (1872), Atchison, Topeka & Santa Fe Railroad Company. Courtesy Kansas State Historical Society.

FIGURE 4.3 Kanza tribal member Jason Murray dances as part of the new arbor dedication, April 25, 2015. Courtesy AP Images/Orlin Wagoner.

FIGURE 4.4 *The Browning of America I* (2000), Jaune Quick-to-See Smith. Courtesy of artist.

FIGURE 4.5 *State Names II* (2000), Jaune Quick-to-See Smith. Courtesy of artist; and Smithsonian American Art Museum, Gift of Elizabeth Ann Dugan and museum purchase.

FIGURE 4.6 *Tribal Map* (2001), Jaune Quick-to-See Smith. Courtesy of artist.

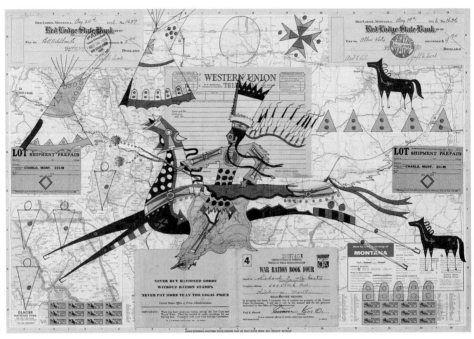

FIGURE 4.7 *Mountain Chief, Blackfeet War Leader* (2008) by Terrance Guardipee. Courtesy Hood Museum of Art, Dartmouth College, Hanover, New Hampshire; purchased through the Virginia and Preston T. Kelsey '58 Fund.

FIGURE 5.1 *Spirit Warriors* (2003), Colleen Cutshall. Little Bighorn Battlefield National Monument. Photo courtesy Bradley Boovy.

FIGURE 5.2 *American Leagues* (1996). Hachivi Edgar Heap of Birds. Courtesy of the artist.

FIGURE 5.3 *Today/Your Host is/Unangax̂* (1988), Hachivi Edgar Heap of Birds. Courtesy of the artist.

FIGURE 5.4 *Reclaim* (1996), Hachivi Edgar Heap of Birds. Courtesy of the artist.

FIGURE 5.5
Gate/Negate (2000),
Bob Haozous.
Courtesy of the
artist. Photo courtesy
Ross Frank.

FIGURE 5.6
*Cultural Crossroads
of the Americas*
(1996) [sans barbed
wire], Bob Haozous.
Courtesy of artist.

FIGURE 5.7 Detail of upper "gate" portion of *Gate/Negate* (2000), Bob Haozous. Courtesy of artist. Photo courtesy Ross Frank.

FIGURE 5.8 Felix River Ranch gate sign. Photo courtesy Daniel Olsen and Henk van Assen.

4

The Art of Native Space

We are "coming back" he says. This is not hyperbole or abstract decolonial audaciousness. On this day, Chris Pappan and a handful of Kanza stand on Kanza land in Kansas. A spiritual leader offers a blessing to dedicate the new dance arbor. A governor speaks. Dance follows. This gathering is the first time the Kanza have stood on and danced on "their" lands in Kansas in 142 years. Pappan describes the moment as a "personal homecoming" (Pappan 2015). Less than two hundred miles separate the Kanza headquarters at Kaw City, Oklahoma, and Council Grove, Kansas. That distance was closed in 2000, when the tribe purchased 168 acres of a previous village site near Council Grove. This site was the tribe's last Kansas homelands before removal to Oklahoma Territory. Yes, the Wind People have come back. Their lands are not gone. Chris Pappan's straight dancers are stepping back onto the map. From paintbrush to footfall. In truth, history shows the Kanza never fully left. Culture tells us they cannot. Today is proof that they never will.

Artist Chris Pappan (Kaw, Osage, Cheyenne River Sioux/mixed-European) tells me the Kanza (or Kaw Nation) are reclaiming their homelands in Kansas. But I am talking to him because he already showed me this reclamation, already painted this scene. Being Kanza under Kansas sky and upon swept earth. His painting *Occupied Territory*, which incorporates a historic advertisement/map from 1872, represents Native space persisting within and emerging from under supposed erasure. Occupied territory. Pushed south. One million five hundred thousand acres subject to Homestead. Thirty thousand free homes. Free homes in Kansas. Waiting for occupants. Lowest special rates for immigrants and their outfit. Visions of White settlement. Kansas is not for Kanzas.

Pappan shares with me that his multimedia piece, *Occupied Territory*, was sparked by the Occupy Movement, which saw protestors nationwide rejecting wealth inequalities produced by our current economic models and global capitalism. Pappan viewed this as a moment to remember an earlier and ongoing occupation: a nationally foundational one that renders all talk of occupation on these lands layered and contradictory in terms of social justice. His creation speaks specifically to the occupation of Kansas. The way his work links to the past, however, is just a prelude to the present and future. Occupation as a verb. Occupation now. From his small creative platform, this Chicago-based artist reasserts a Kanza powwow scene on a historic 1872 advertisement/map once used to facilitate and mark the unmaking of Kanza homelands. Pappan notes this unmaking and then presents a "forward-looking" sense of Native possibilities. This same map now helps to remake those homelands. It reflects the remaking already at work. Dancing. Foot on land. On land.

The Return of the Kanzas

As I look for a better resolution of the original 1872 advertisement/map, I hear a train's horn call out and the low, almost unnoticeable hum of the wheels on the rails. Perhaps it is fitting that as I look over a copy of the historic document a train rumbles through my university campus, audible even inside our library stacks. The map portion of this document marks the starting point of the Santa Fe Trail. It announces the birth of the Atchison, Topeka and Santa Fe Railway. I can easily trace the shaded river-like pathway laid out by the AT&SF mapmakers and marketers. They are selling Kansas. The train is both a metaphor and a vehicle for dispossession. Entryways into Kanza lands. Starting points for unmaking Kanza lands. Council Grove sits at the mouth of this swollen iron river.

Pappan's art draws direct attention to how the railroad marked off new space via this map. I search the map for Satanta (see chapter 3). It will not be established until forty years later. Satanta is not there yet. Set-tainte still is. He is alive, although recently imprisoned in Texas for a wagon train raid. Soon he will be dead, buried by this newly forming space, drowned by this river of steel, by the inundation of new bodies, cultures, and maps. *Occupied Territory* reminds us how maps, along with brute force and policies of invasion and removal, reenvisioned and then constructed this land anew. Pappan's remapping project demystifies the materialization of that spatial project. It chronicles the unmaking—and now the remaking. The year before this original 1872 document was created, a federal law relocated the Kanza to Oklahoma Territory, fearing the constant reprisals and incursions of White settlers. Citizens and lawmakers feared Set tainte. Pappan's production encapsulates the simple irony of settlers begrudging the indigenous peoples, demanding their removal—the now obvious (at least to most) injustice of a collective action carrying out simultaneous assumptions of superiority, "safety," and removal. On this 1872 document, the railroad company offers White and Black settlers land twenty miles to either side of their main line through the state, deeded by federal land grants (fig. 4.2).

Viewed through Pappan's painted lens, the original 1872 advertisement/map belies colonial intentions. The Pottawattamie Reserve and an amorphous Osage Indian Trust Land are still present, labeled even. It seems that Native space is persisting within the homesteading map as well as beyond it. The map itself uneasily represents the ongoing effort to change the geography. Pressing against indigenous persistence are referential—perhaps ventriloquistic—spaces attempting to unmake Native space: the town of Hiawatha, counties called Ottawa, Cherokee, Comanche, Osage, and Kiowa. Kansas.

Just as in Satanta, however, places like Council Grove offer a story not just of contested space, but also of potential for shared or comfortably overlapping space. Today, the people of Council Grove invite a new relationship with the Kanza Nation. In 2001, Ron Parks, historian and one-time director of the Kaw Mission State Historic Site, noted that a tree-planting project would serve to remember the Kaw and to "cement relations" with the Kaw Nation. While some of these efforts may simply work to give or recall some "ethnic flavor" for their town and garner a new tourist angle, this exchange rests more firmly in Kanza hands than ever before. The town raised funds and volunteers to plant 807 trees, a number purposely selected for each one for the 807 tribal members listed on the census prior to their removal from Kansas. An opening gesture. A material and spatial commitment. Re-placement rather than dis-placement. A possible template useful for whatever lands "we" all stand on.

At the same time as this cultural-historic tree project, the Kanza are working to have the land they purchased placed on the National Register of Historic Places. It has already been established as Allegawaho Memorial Heritage Park to tell the story of the Kanza and their last Kansas leader. The Kaw Nation's park director, Betty Durkee, noted at the time that "this officially marks the return of the Kanzas to the state that bears their name" (Anon. 2002). Durkee's proclamation is decidedly spatial. People and place, and mutually defined. This purchase visibly and concretely reestablishes Kanza physical and geographic presence.

But, the Wind People never completely left Kansas. Homelands travel with us in memory and identity, in story and ceremony. More concretely, a quick historical survey finds that they physically returned for the dedication of the "Unknown Indian Monument" in 1925 (Fry 1990). On August 12 of that year, twenty-five mounted Kanza warriors filled a ceremonial procession that carried the body of a Kanza man being reinterred after his body was discovered by a sympathetic town leader. Roy Taylor (identified as Pete Taylor on the Kaw Nation website), a Kanza leader and grandson of the last Kansas leader, Allegawaho, said through an interpreter that he was "glad again to be in the land of my fathers" (Fry 1990, 23). Vice president of the United States Charles Curtis, a Kanza tribal member who spent much of his youth in this very land, arrived at Council Grove in 1930 to officiate yet another dedication for that same monument (he missed the first while serving as a senator).

The Kanza returned to Kansas for an unknown Kanza man. They returned in the form of the highest-ranking Kanza in the American nation. They returned for ceremony and culture. They returned for identity. They returned for their lands.

Perhaps it appears a simple action to merely paint over a document. But Pappan's is no simple act. The paint is not merely placed. Nor is the choice of this document incidental. Looking at the images together —the original 1872 advertisement/map and Pappan's multimedia creation based on it—changes the original. We know that all engineering requires constant maintenance or it falls to dust. Merging with his paint and vision, the original map now frames the social, cultural, economic, political, and spatial engineering feats required to produce an indigenous absence. Rather than serving maintenance of that long-dominant construction, Pappan conducts a precision demolition. Absence becomes central to the construct. Removal is revealed as narrative and material. Disruption. Space is unmade. Space is remade. Council Grove is being transformed again.

Mobility and Indigeneity

Differing deployments of mobility permeate spaces like Council Grove, Kansas. But Council Grove is not unique in this tense relationship between mobility and indigenous geographies. In general, mobility is the capacity to enact movement or to deny stasis. This capacity is physical in terms of bodily movement from one part of the world to another. It is also cultural, in that social practices and their connections to identity are shifted and adjusted in response to new experiences and new ideas coming from within or from without. Mobility is also restricted by physical, psychological, and social obstacles, and the interrelations between these factors. In this light, we can see that colonization and dispossession are about regulating capacities for enacting mobility and space. So, I want to take a moment to draw attention to the connections between mobility and place in Native epistemologies, identities, and ontologies. I am interested in this connection because it simultaneously privileges indigenous spatialities and highlights the geographic challenges of a neocolonial world. In this way, I want to build on the interventions by Epeli Hau'ofa's reframing of land, scale, and mobility for Pacific Islander peoples and to consider how artists use maps to revision and recover indigenous geographies (Hau'ofa 2008).

Maps have long been tools of colonization and mechanisms for the construction of particular kinds of space, both conceptually and materially. More recently, maps have also been redeployed to subvert colonial outcomes. They have served to restore tribal geographies and chart culturally significant relationships to the land, often with limited access granted only to those deemed legitimately entitled. These "subversive" remappings—including those I trace in the work of the artists discussed in this chapter—and the implied tensions over space in relation to Native peoples and colonized lands are not unexpected nor simply reactionary. As the late and esteemed Dakota scholar Vine Deloria keenly reminds us, Native worldviews are rooted in spatial frameworks, whereas Western worldviews are organized by and operate through temporal structures (Deloria 1994).

This in part explains the significance of place to Native cultures and identities, which are fundamentally embedded within specific geographic locations. Consider the Diné and Diné-Tah, their place of creation or emergence, which signals more than just a history of residence or occupation. Indeed, both the anthropological research and tribal oral histories tell a migration story of the Diné from the north to Diné-Tah. Despite this process of movement into what is now called the Southwest, Diné-Tah is the Diné homeland. Diné-Tah is the location for Diné ceremony, where the stories originate and where the umbilical is earthed. Diné-Tah both initially produced and continually produces the Diné, and thus they cannot both leave and remain Diné, at least as a collective identity.

Yet, this place-ness does not negate mobility, even if it represents a mechanism for internal and physical constraints. It was created through mobility and is now comprehensive and secure enough to sustain some degree of continuing mobility. As my Hopi, Pueblo, and Zuni friends and I often joked during my time in New Mexico, Navajos seem to be everywhere—evidence of their extensive mobility! We see similar examples outlined in the work of Myla Carpio discussing the establishment of Laguna Pueblo colonies across the Southwest (Carpio 2004).

The American colonial project was effectively centered on encouraging particular kinds of Native mobility (relocation, land cession, assimilation) while discouraging other kinds (use rights, internal trade routes, counter-colonial alliances, revitalization movements, sovereignty) all the while using this complex of interventions to subsidize and facilitate broader European American capacities for and possibilities of mobility. In short, Native people were moved away or out of the way so White people could move around and settle down. Indeed, settler space is secured and defined only when indigenous mobility is denied, and only through this denial. This imposed stasis works to strangle and submerge Native space, although as I argue below and throughout this book, the violence of settler spatialities does not always succeed in the suffocation or the submersion of indigenous geographies. This is because space is not merely a set of abstract coordinates that can be claimed or parceled out, but is also sets of meanings generated and sustained by the relationships between material and cultural worlds. While bodies may be moved and abused, and lands given new shapes and meanings, Native space is not easily dispatched.

Some might ask how this argument about place-centered identities and Native space can account for the mobility and changes during the precontact periods, how this can reconcile the innumerable instances of conflict and movement that took place in tribal life before Europeans ever established a dominating presence on the continent. It would indeed be naïve and uncritical for me or anyone else to suggest that each and every Native nation is uncontextually tied to the very specific land base where they resided at the moment of European contact, as this would erroneously leave Native peoples without room for creativity, flexibility, and mobility. We know that all Native people moved and sometimes conflicted with one another over territorial claims or at the very least over rights of access and control if not typically outright possession. Further, many tribal identities have been explicitly consolidated by such movement and transformation, formed when a people "emerged" after travelling in search of their homeland or changing how they operated in the world after being imparted wise teachings.

The stories of connection to land are not just observations but actions, or relationships marking changes in mobility. They narrate the move from being a stranger to belonging. I have already mentioned the Diné migrations. We can also consider the movement of the Anishinaabeg, who travelled from water to water, leaving the Atlantic coast to relocate in the Great Lakes region following a prophecy and a shell that led them to Anishinaabe Akiing. Another example might be the Iroquois Confederacy, the very founding of which was centered on disrupting cycles of violence and warfare and promoting new relations and ethical, social, and political protocols among the peoples who would become the unified Five Nations. Once it was widely presented and accepted, the Hiawatha wampum belt abated land incursions and revenge plots. At that moment, a new, Haudenosaunee geography was articulated and mutually recognized. They created something new. Like the Diné. Like the Anishinaabeg. They changed through mobility.

This tension between mobility and homeland is interesting precisely because it constitutes what it means to be indigenous in the current era. Creativity, flexibility, and mobility are essential characteristics of a current Native life also intent on recovery, revitalization, tradition, and the rights of sovereignty (or the equivalent in those places where such a legal recognition is denied). How else can the removed tribes now centered in Oklahoma make sense of their indigeneity? As my brief examples suggest, movement and homeland are only in opposition if one artificially imposes the impossible notion of perfect stability and timelessness. Spatiality reduced to place-ness.

So far I have explained how the movement and transformation of Native peoples is the means by which they came to be, the way they created spaces of home and how those spaces of home in turn created them. This process of becoming did not, however, happen at the instant that a group of people decided to build homes and plant crops and tell stories about their arrival. This would mimic the colonial model of belonging, which confuses occupation and labor with connection and relationship.

Consider the ridiculous example of the Spanish conquistador Hernán Cortés reading the Requerimiento upon landfall and proclaiming land rights and the absolute subjecthood of local Native peoples to the crown and the pope before helping initiate a continental transformation. Think of those original Jamestown colonists who refused the advice of the local Powhatan Confederacy tribes and established their homes on a known floodplain and swampland before facing ecological-based hardships and ultimately "disappearing" from the land. Just as audacious were the Boston colonists who lazily dressed as "Mohawk" warriors in order to throw tea into the harbor, partly in response to taxation but mostly in protest of British Crown policies limiting colonization and occupation of sup-posedly wild interior Native lands that they wished to remake into recognizable, submissive geographies.

In contrast, Native groups became Native by hard trial and error, experiences earned well beyond the instantaneous opportunity of warfare or emigration, of proclamation or simulation. Being Native to a place, as Daniel Wildcat and Raymond Pierotti remind us, depends on extended periods of time by which a people can observe and study their surroundings, learning from the land-scape, plants, animals, and weather until they have effectively come to find the appropriate human role within and its balance with the ecosystem (Pierotti and Wildcat 2002). This is Native space before settler colonialism. A traditional model of Native space is produced by a relationship not dominated by human self-interests. Barring this, human occupants are just—as most Native creation stories testify—dangerously insecure beings in uneasy tension with their earthly surroundings.

So, postcontact, when Native peoples were indirectly pushed or physically relocated outside of their precontact locations, their ability to establish a com-prehensive indigeneity was compromised, but not abolished. Despite their com-pelled movements, they retained the knowledge base and skill set to reestablish a Native space should adequate mobility be allowed, if they were so inclined in this context, and if they possessed the proper patience. They could re-locate their place in the world through observation and determination, but not through the "hard work" and/of transformation favored by colonists, which required chang-ing the land to match a predetermined, human-centered template. The Kanza were deeply engaged in this process, having moved from the east centuries before, being removed in 1873, and having to begin again in Indian Territory. Likewise, the Kiowa, as discussed in the previous chapter. The key distinction in a colonial context, then, is that the time lines required for such transitions and transformations are violently disrupted, and in unprecedentedly comprehensive fashion. Generational projects often become impractical when immediate survival is not guaranteed. Thus, the Oklahoma Territory experience for the Kanza and many others can be seen as an ongoing project of Native space constrained by postcolonial limitations, further complicated by tribal returns to pre-removal homelands or new efforts to extend beyond what are deemed "reasonable territories" by the nation-state.

Maps of Recognition

Let's return to maps, and art. Maps are complicated. They are political documents. They are simultaneously fluid and static. Both artistic and documentary. They are reflections of the relationship between space and identity—of spatiality. Given these characteristics and this complexity, they offer a great deal of conceptual material for an artist. In light of my discussion of mobility, I want to return to contemporary Native artists. In *Occupied Territory*, Chris Pappan painted over a historic advertisement that was also a map. Pappan's use of map and sign parallels a growing body of Native artwork and artists keen on how these devices offer them and their audiences a visual opportunity to reconsider space, to recognize and restore Native mobilities. These artists specifically redeploy maps to establish recognition of indigenous spatiality—a unique formation of relationship-conscious interaction in the co-constitution of people and place. These artists are actively constructing and participating in ongoing indigenous geographies in both mundane and sometimes explicit or spectacular ways. They represent these spatialities and help reproduce them. Using differing techniques, each uses maps to speak a profound commitment to and recognition of Native spaces as historic and ongoing realities that simultaneously refuse, resist, and navigate imposed colonial geographies. They also visually convey the ways that mobility reflects a flexible understanding of space and how the relationships that generate space and identity can be and are configured and reconfigured.

Since I have already discussed Chris Pappan's work, I now want to add selections from a wide-ranging series of maps by Jaune Quick-to-See Smith (Flathead/Cree/Shoshone), and then focus on one specific collage piece by Blackfeet artist Terrance Guardipee. I start with Jaune Quick-to-See Smith, as she offers a relatively "pure" use of maps in that her series centralizes the map as form or container that can be emptied and then refilled (Rader 2011, 53). While many of her other works embed a greater density of tribe-specific content, these map pieces tend toward a generalized and overarching treatment of Native space, or a pan-Indian geography. As her cartographic and cultural representation is broad, her approach intentionally targets shared indigenous spatialities. In preparation for the map series, and while brainstorming some collaborative work for the US quincentennial with other indigenous artists, Smith imagined how best to center Native people and issues. Before landing on (and ultimately curating) a "subversive" gallery showing on Christopher Columbus (called Sublomuc), they considered planting a flag and claiming Spain, or "going to Missoula and changing all the street names to Indian names" (Sasse and Smith 2004, 20, n22). These options nicely outline her spatial frame. All three options—flag-planting, street sign renaming, and the art show—like her map art, center on producing spatial disruptions and recentering what might be seen as subversive (or at least buried) Native geographies.

In the same way that Smith would have replaced Missoula's street names (a serendipitous segue from my previous chapters), her map series repeatedly wipes clean identifying markers of the US American nation and states in order to stake out an underlying and ongoing indigenous presence. Smith redeploys Jasper Johns's famous approach to a series of paintings named *Map* (in 1961 and 1962) and *Two Maps I* and *Two Maps II* (both in 1966), in which he progressively moves from splashy colors to gray and then all-white and doubled mock-ups of the ubiquitous American states maps. Johns purposely sought to renew and revisit an image that had become mundane. He took the ubiquitous multicolored map found in classrooms (and beyond), eschewed the clean lines and rendered it "messy" nearly to the point of near-non-recognition (perhaps saved only by retaining state name labels). He offers an impressionistic interpretation of an already symbolic device that had begun to become invisible. For all its social and visual impact and commentary, however, the content held no further intentionality beyond the act of looking anew and asking viewers to linger long enough to center the viewing moment and to stretch the aesthetic experience. Johns changes the map, but not necessarily its meaning.

Smith's maps invite us to look anew as well, but then prompts the viewer to shift perspective on what is being viewed. In many ways, her impressionistic interpretation intentionally increases the invisibility of some elements of the standard map. She consciously uses a "seduction" technique—what she also refers to as a "sneak up"—that encourages her viewers to look closer and then "reach a level of understanding that her messages have serious implications" (Sasse and Smith

2004, 13). Her engagement with the audience starts with removals and additions. When Smith constructs her maps, for example, she often adds words and text, but consciously "bury[s] them in the paint, hiding them here and there so they crop up, although you wouldn't discover the words until you came up close" (Abbott 1994, 218). By embedding layers of engagement, she articulates her approach as rejecting "what you see is what you get" and based on a belief that "it has to be provocative in order to bridge the gap in understanding between Indian issues and the mainstream" (Abbott 1994, 218–219). In an interview with scholar Lawrence Abbott, Smith explains that when making art she is "not really thinking about how to be accessible but rather how to communicate to the viewer in layers" (Abbott 1994, 219). Lucy Lippard, an art scholar and longtime friend and collaborator with Smith, points out that in her public installation work—and I would argue her map-based work as well—she "wants to create spaces that *haunt* people within which she can transform the ways they think, giving them information they simply can't shake when they leave" (Lippard 2004, 92; emphasis added). Smith herself indicates the intention for haunting is to reach beyond just intellectual comprehension and to create "something that will stick to you, body and soul" (Lippard 2004, 92).

In her capable hands, the US map becomes an entry point for cultural, historic, and political engagement. It becomes an interrogation of space, of the ways we construct and sustain spatialities and of the fact that they too exist in layers. Maps become documents of what existed in the past, but also what now exists, and how we know about these existences and transitions. Smith tackled American state maps as a form beginning in the year of the US quincentenary, a moment of national commemoration that threatened to rhetorically reduce the multiplicity of spaces and spatialities to a singular space and a particular (colonial) spatial process. She ultimately produced *Indian Map* (1992), *Indian Country Today* (1996), *Memory Map* (2000), *The Browning of America I* and *The Browning of America II* (both 2000), *State Names I* and *State Names II* (both 2000), and *Tribal Map I* and *Tribal Map II* (both 2001). As the list suggests, a single entry point to indigenous geographies (and histories and cultures) is not sufficient. She acknowledges multiple entries and multiple possibilities in order to communicate with audiences, gain traction with her interventions, and account for the multiplicity of indigeneity. As Dean Rader astutely observes, "So complex is our visual and cultural notion of Indian space that one map is not enough. Smith must paint many" (Rader 2011, 71).

In *Engaged Resistance*, Rader offers an overview and comprehensive treatment of Jaune Quick-to-See Smith's map works. He illustrates the map series' value, placed in context with other forms of art and artistic practices, as an act of indigenous cultural resistance and part of a larger "semiotics of sovereignty" (Rader 2011, 53). Given his solid presentation, I want to turn an eye toward further articulating and emphasizing the *spatiality* embedded within Smith's map creations, which seems rich with possibility given her use of the map-as-device and does not figure strongly in any analyses or discussions of her work.

Treating her maps as a spatial text, then, we should particularly note how the work sustains a contemporaneous impact. I start with *The Browning of America I* and *The Browning of America II*. These paired pieces centralize a broadened (pan-Indian) indigenous cartography by largely erasing state boundaries and placing European invaders as a kind of sidebar. The south-to-north paint drip technique "browning" the states to the point of erasure suggests a temporal reversal toward a more ancient geography, as well as a general tension over racialized landscapes. But this is not, therefore, simply a move toward the past before European arrival. These remappings speak as much to the present as the past. Smith's displacement of the US nation and of historical agents of colonialism is a reordering of overlapping indigenous and European-American geographies. Hers is an articulation of colonialism as process, not event (Wolfe 2006); it also does not provide any discrete or viable point on a time line to which one could imagine returning.

The active participle of brown*ing* can easily indicate present or future tense, and thereby eliminate a purely historical lens. This reading is supported by Smith's inclusion of the list of "invaders from the east." The invaders are listed using historic terms, but she includes a 1,001-year time line that stretches from the year 1000 to the year 2001, a future moment one year beyond the piece's creation in 2000. These elements resist an invitation to read *The Browning of America I* as primarily remembering past geographies. The use of petroglyphic figures understandably leads some to this connotation. Rader, for example, reads it as reminding viewers that "before states, before state names, before borders, and before federalism, this space was inhabited, inscribed, and marked by people who remain part of and parcel of its ontology" (Rader 2011, 66). In this articulation, while the Native people exist and remain "part and parcel" of the nation-state, Native space is conceded as "before." But petroglyph symbols are not

bound by time, despite Western tendencies to render such cultural tools anachronistic with the contemporary moment. In his one-sentence prologue introduction to the chapter focusing on Smith's work, Rader states that he "looks at
the way Smith uses the iconography and semiology of maps to remind the viewer
of land reclamation, broken treaties, name changes, and relocation" (Rader 2011,
3). Although land reclamation is an undoubtedly present set of activities, the
other "reminders" seem largely presented as historic processes (although with
ongoing consequences) and still center the US nation-state.

In Rader's setup, Smith's iconography supposedly replaces those non-Native
maps she uses within her work. We are left with a Eurocentric chronicle of historic processes that have fallen out of collective (non-Native) memory. I would
argue that this kind of reading slips toward rendering Smith's resistant interventions as little more than revisionist narratives. It also presumes a zero-sum spatial game that cannot sustain multiple and overlapping spatialities. If we instead
insist on observing how the map conveys a contemporary presence and futurism,
we might more easily read the images and title of the painting marking Native
space as present. Otherwise, we risk continuing the erasure of Native space that
dominant maps already imply by restricting them to the past. As Smith notes,

> Ours are the only religions that directly name the plants and animals and insects
> here. Ours are the only religions that map our geography—such as our medicine
> tree down in the Bitterroots or the sleeping giant. All our morality teaching stories
> come from our land. I can't say strongly enough that my maps are about stolen
> lands, our very heritage, our cultures, our worldview, our being. . . . Every map is a
> political map and tells a story—that we are alive everywhere across this nation and
> are in residence everywhere whether we are recognized . . . or fighting for recogni-
> tion. (Sasse and Smith 2004, 8)

Using a notable present tense, Smith lays out a lively and comprehensive indigenous spatiality that struggles with recognition not just in the political sense
but in all the ways of "our being." This also matches other indigenous mapping
projects aimed at intervening nation-state erasures of Native peoples and spaces.
As Latin American geographer Sarah Radcliffe notes, although in reference to
strictly cartographic projects, maps are always "a means to articulate notions
of different futures articulated in contradistinction to state multiculturalism"
(2011, 140). Further, the fact that maps such as Quick-to-See Smith's represent
"pedagogic (not merely resistance) projects" clearly indicates a future-oriented

vision aimed at changing, decolonizing, future knowledge, materiality, and practice—in short, a practiced and lived spatiality (143).

We see a similar approach in Smith's *State Names* pairing, where state and nation lines are now the palimpsest rather than the overlay. The maps in *State Names I* and *State Names II* appear at first glance to be messily rendered but otherwise standard maps. Closer inspection then reveals that some of the state names are missing. While state names within the national borders are "missing," some state names outside the national borders are not, suggesting that the map offers a representation of something other than those nation-states. The absence of some US state names ultimately helps highlight the presence of state names in Mexico and Canada. With sufficient attention, or contextual guidance, the viewer is encouraged to experience a third encounter in which she realizes that the only remaining state names are those derived directly from indigenous languages, heritage names adopted by Europeans and Americans. The move across current national borders likewise presents a broadly, in this case hemispheric, indigenous map. Mvskoke poet and artist Joy Harjo suggests this technique helps us see that in the context of modern nation-states "the boundaries are false and they drip in shame" as well as reminds us that "words are powerful and create the reality of the world in which we live" (Harjo 2004, 66). Smith's exclusionary mapping of only indigenous-based names documents that these names continue to exist at the same time that it memorializes that even this list of names is incomplete.

By their unmarking, the unnamed spaces actually become marked as places of active and ongoing erasure. Absence is centered and questioned. She brings attention to the process of transformation of the land and the act of creation required for changing place-names. The savvy viewer might come to recognize that even the marked spaces, as Harjo also implies, are incompletely mapped. Indigenous names and spaces do in fact continue to exist, and thus a viewer might finally comprehend or at least wonder about the continuation of those other geographies, other maps.

Smith's strategic use of addition, removal, and layers continues in *Tribal Map I* and *Tribal Map II*. In these pieces she removes all the state names, although she redraws their boundary lines relatively unaltered. Instead, the state outlines now contain what appear to be cut-and-pasted lists. Looking more closely, a viewer finds names of tribal nations placed in their homelands or places of relocation (such as Oklahoma). What was mostly implied in the *State Names* pair is here rendered explicit. The state names are now replaced by tribe-specific names and locations. The complexities of Native geographies start to take form and are gaining increased if still simplified precedence over the state lines and national image. Her technique, though, suggests that Smith is still not able to reconcile the totalizing tendency of the map with the diversity, fluidity, and multiplicity of indigenous spaces. Noting the intentional messiness of her map, Harjo suggests that Smith recognizes that even her revised map can have "no easy boundaries or lines." Harjo appreciates the need for messy lines here, precisely when trying to chart Native spatialities. Harjo rhetorically, perhaps sympathetically, asks, "How do you concurrently illustrate the original concept of sovereign tribal nations who held lands in common and overlap from other tribal nations?" (Harjo 2004, 65–66). This articulation highlights the challenge of representing a holistic Native spatiality using a technology primarily developed and useful for delineating European concepts of property and navigating unfamiliar lands.

In treating these paired pieces, Rader asks an intriguing question: "Why create a map that erases the names imposed by the government but does not erase the borders established by the same government?" (Rader 2011, 58). He suggests that by replacing state names with "the various names of its indigenous tribes" the land would become a "different entity, not simply a differently named space" (Rader 2011, 58). He continues: "Even more provocative is the possibility of doing away with states altogether [as in *The Browning of America*] and conceptualizing the land we think of as America only as a collection of Native tribes" (Rader 2011, 58). While his thought exercise is intriguing and a useful heuristic strategy, it also assumes that the lands of Arizona, for example, cannot and do not already exist as both Arizona (the state) and as a different entity comprising indigenous names and spaces. While he rightly notes that this layering tension "comes out of Smith's desire to both engage and resist," his suggestion does not fully account for the now deeply intertwined nature of indigenous and settler colonial spaces. Extending the spatially informed intervention made by Smith to the point of eliminating state lines or the nation altogether ultimately ignores the realities and tensions of overlapping spaces—tensions expertly represented by Smith and Chris Pappan, as well as by Terrance Guardipee, who I turn to next.

Artifacts and Presence

Where Jaune Quick-to-See Smith removes colonial overlays to ostensibly present underlying indigenous geographies, Terrance Guardipee leaves the colonial maps intact and instead imposes a new Native overlay. Both approaches seek to establish that Native space persists in tension with the US nation-state, just as tribal cultures continue in tension with non-Native cultures. In fact, space and culture—the interaction between which I discuss as spatiality—are co-constitutive. This is also true, it should be noted, about the relationships between indigenous and Western spatialities. Each is partly created and informed by the other, and thus attention to one is attention to both. Despite the cultural-spatial dialectic, many scholars have ignored these connections and tensions in Native art and beyond. Guardipee turns his artistic attention precisely to these tensions and focuses on the Blackfeet rather than larger pan-Indian geographies. In *Mountain Chief, Blackfeet War Leader* (2008) in particular, spatiality takes center stage as Guardipee stamps colorful drawings of nineteenth-century Blackfeet cultural life over a contemporary map of Montana. I focus on this single piece, which I find exemplifies Guardipee's intriguing "map collage" technique, in which he layers artifacts and ledger-style illustration over maps.

Guardipee was one of the first to use the neo-ledger style now practiced by a number of contemporary Native artists, including Chris Pappan. The original ledger artists seamlessly expanded a Plains Indian tradition of visual narrative drawing and painting previously found as pictographs on rocks, bison hide and tipi painting, and war shield decoration. Ledger book artists emerged during the late nineteenth century, when imprisoned Plains warriors continued their illustrative storytelling practices on the only drawing surfaces made available to them once bison hides were rendered unavailable by the planned destruction of the herds. These artists picked up discarded or surplus trading post sheets and accounting ledgers to draw and paint their stories and memories.

Neo-ledger artists like Guardipee make explicit tribute to the original ledger artists, although they are distinguishable by their incorporation of new elements (such as layered artifacts), their transplantation of the ledger styles to intentionally selected mediums (such as historic maps or site-specific ledgers), and, often, for making expressly cultural or political statements (critiques of assimilation or colonialism). Guardipee's map collages routinely comprise three layers: a late-twentieth-century map of Montana, early-twentieth-century legal-commercial-industrial documents, and brightly colored ledger-style illustrations. Like the original ledger art, they all feature real individuals and stories. The original ledgers, ostensibly just representing events, culture, and histories, also inevitably contain suggestive scenes of or allusions to the processes of dispossession, cultural transformation, and confinement. The art form itself, if nothing else, certainly belies the tense engagements between tribal peoples and Americans.

In *Mountain Chief, Blackfeet War Leader* and many of his other map collages, Guardipee draws attention to the colonial and geographic tensions implicit within the original ledger art genre. Guardipee's merger of ledger aesthetics and maps offers viewers an intentionally spatial frame of reference, and then opens a small window onto how that space is produced by Blackfeet people and culture. Guardipee's need to explain the symbols he uses emphasizes the cultural and spatial ambiguities and fluidities presented within his work. This is a necessary process, as he tries to change how many viewers engage with "Montana." For non-Native viewers (if I may simplify for a moment), it questions the meaning of that geography. What is Montana exactly? What is given importance? Who decides? For Blackfeet or other Native viewers, his images reaffirm and give visual record to the layered spaces within which they operate. How do I navigate the state? What sites are important to us? How does this differ from dominant values?

So, what precisely does Terrance Guardipee give us to read? Let us first take a look just at the rich ledger illustration overlay. *Mountain Chief, Blackfeet War Leader* places the central human figure, a real person of importance to the Blackfeet named Mountain Chief, in the midst of rather glorious action. He, with his horse, strides over the map while his paraphernalia dramatically floats in mid-bounce. The war leader flies across and largely envelops via his scale the landscape of Montana, which is covered by markers of indigenous culture, presence, and prosperity, but also embellished with artifacts of colonial encounter. Above Mountain Chief we find the sun and a symbol for the Blackfeet cultural hero Morning Star. Behind the rider we find five mountains, each marked with a stacked pair of stars, also referred to by the Blackfeet as "puff balls" (Anon. 2004). John Ewers notes that puff balls illustrated on mountains in this way represent falling stars, or stars that have come to rest on earth, thus granting them special terrestrial-celestial significance (Ewers et al. 1976).

Above and below the mountains, spirit or sacred horse figures repose, seemingly observing the galloping warrior and his warhorse. In the direction of the rider's path we find four male figures and two decorated tipis. The male figures are abstracted, consisting of downward-pointing triangle bodies, simple circle heads, and red dots at the heart point. Mountain Chief and his mount are richly decorated and outfitted. Mountain Chief carries a lance and staff and sports the telltale cylindrical war bonnet unique to the Blackfeet. Like the tipis, his buckskin clothing is decorated by a cluster of stars, but also features butterfly symbols on his legs, arms, chest, and headband. The horse decorations match the rider's design of stars, butterflies, and Morning Star. The tipis are both decorated in a traditional Blackfeet, banded design (Ewers et al. 1976). On the upper "sky band" opposite the entryway, Guardipee depicts a symbol of Morning Star, and star patterns on the smoke flaps. Resting along the lower or "earth band," both tipis feature common Blackfeet illustrations of triangular mountain symbols and rounded hill symbols. The central or "all creatures band" remains unadorned. In short, an entire cosmology represented.

If we take a closer look at the backdrop and artifacts of *Mountain Chief, Blackfeet War Leader*, we also find that Guardipee creates his collage over a 1971 Texaco-produced roadmap intended for automobile drivers and gasoline customers. Guardipee includes a 1943 war ration booklet, ration stamps, American Railway Express government carrier receipts, 1926 Red Lodge State bank checks, and a Western Union telegram form, all of which reveal landscapes of commerce, nationalism, and (neo)colonialism. Given the historic precedents of the Blackfeet, some of which I outline below, we begin to see that the artifacts Guardipee affixes to the map must be read as tools of nation-building, historic mechanisms by which the United States actively reimagined and reordered Blackfeet lands as White (or perhaps in contemporary times, multicultural) American space. This stands in contrast to the ledger illustrations, which can be read as part of a Blackfeet nation-building. Both the artifacts and the ledger illustrations point to the context through which Blackfeet have sustained their cultures, identities, and geographies. In these ways, despite the tribal specificity, his example conveys most tribal peoples' experiences with dislocation, dispossession, and various mechanisms of containment and regulation.

The contrasts among Guardipee's collage's layers highlight competing, intertwining, and overlapping spatialities. These contrasts seem to signal and return us to the importance of mobility: modes of mobility, structures of mobility, mobility as mechanism for domination, and mobility that defies confinement. It also suggests that this vital characteristic is, simultaneously, built on the constraint of Native mobility. At the same time, Guardipee's artwork emphasizes flexibility and adaptation. The Blackfeet images and symbols are important cultural markers that sustain contemporary identity and space. They offer a glimpse into a Blackfeet understanding of their relation to the world. It signals their spatiality, the core of a unique set of connections between a specific land, identity, and culture.

Guardipee clearly rejects recurring expectations of Native cultural, economic, and political stagnation (or "backwardness") even as he celebrates and protects traditional cultures and identities. Rianna Hidalgo confirms that Guardipee constructs the layers of his map collages as doing different kinds of work. The map backgrounds "tell a story of a shifting and changing Montana"; the ledger-illustration overlay, on the other hand, "represents his tribe's people holding onto their culture and identity" (Hidalgo 2015). "Even though these changes come," Guardipee says, "we remembered who we are and where we came from" (Hidalgo 2015). In an interview with a regional railroad magazine, Guardipee explains that he wants his art to help "keep the spirit of the Blackfeet alive" (Pfeuffer n.d.). In another interview, he indicates that the symbols of "culture, society and mythology" within his creations act as "a way of ensuring they are never forgotten" (Hidalgo 2015). Hidalgo points out that Guardipee worked hard "convincing elders that telling a story of Blackfeet strength, spirit, and beauty would be good for his tribe" (Hidalgo 2015). The elders, she says, "didn't always welcome sharing Blackfeet imagery with the world." This point of discussion between the artist and tribal elders strongly indicates the degree to which the traditions and symbols he presents with his art are ongoing cultural components worthy of protection. Were they relics of the past, protection would be less urgent, or unnecessary.

Guardipee's technique of overlaying Blackfeet illustrations on roadmaps, then, parallels Blackfeet historical and spatial realities that undergird dominant spaces and settler colonial geographic narratives. Despite the layered contradictions and spatial power fused into the landscape, Guardipee does not attempt to erase the Montana map, any more than the Blackfeet attempt to extinguish the non-Native population. The contemporary Native spatialities presented by Guardipee, Smith, and Pappan are properly shown operating alongside (as well as against, with, and within) the contours of non-Native power and geographies. This aligns well with Jaune Quick-to-See Smith's practice of leaving state lines and/or national borders intact (however roughly rendered), and with the Kanza efforts to craft new relations with the people of Council Grove, just as they reestablish a Kanza spatiality on the site of their removal. This is what remains a possibility in Satanta, and what is missing entirely in the hundreds of Indian Villages across the country (as discussed in the previous chapters).

Reading Landscapes

Guardipee's background map features a legend titled "how to read your map of Montana." His choice of maps either suggests the artist's wry sense of humor, aimed at drawing his viewers' attention to the productive act of reading and mapping, or presents a serendipitous coalescence on this same point. In either case, the legend reveals the productive and destructive aspects of mapping and its corresponding crafting of material space (Harley 2001). This "guide" proposes to instruct its viewer "how to read." Yet this boxed set of instructions clearly already requires a technical reading proficiency. It requires shared points of interest and shared spatial understandings. Noticing this expected cultural fluency can open us to a larger discussion of how producing this map, as well as its directions for reading, is replete with intention, value, and consequences, all of which must be subject to interrogation. We can see that the map features are decidedly not random or neutral. Thus, if we prove to be "bad" readers or intentionally transgress in our reading, such a map can point to critical cartographic approaches and reveal the collaborations between mapmaking and empire-building. Indeed, one would be hard-pressed to understand the complexity of Guardipee's collage without wielding such an alternative reading skill.

How does Guardipee and how do the Blackfeet read Montana? This is a complex question with numerous answers, but in order to further contextualize his map collage, I want to highlight how Guardipee's work shares several Blackfeet spatialities and reflects the ways Blackfeet geographies have been affected and shaped by the experiences of settler colonialism and neocolonialism. Consider, for example, that in 1915 a delegation of Blackfeet travelled to Washington, DC, and "protested renaming lakes, mountains, glaciers, and rivers in [Glacier National] park" (Keller and Turek 1998). They were concerned that Americans were increasingly and rapidly dispatching with the traditional Blackfeet names and all of their attendant meanings and stories. The capital delegation recognized that the acts of renaming constituted an especially insidious effort to produce a vastly different and non-Blackfeet geography where it had not previously existed, an effort already supported by American control over the land, forced land cessions, cultural impositions of "civilization," and restrictions on tribal hunting and fishing rights. While American narratives always prefigure this transformation and require this spatially significant act of naming/claiming, the Blackfeet stood at this moment of transformation actively seeking its disruption even after what was presumed a concluded conquest. In fact, too many Blackfeet individuals still remembered and were still defined by their tribal geographies. Their spatiality was active and persisting. They sought to ensure it continued that way.

Contemporary European American author-explorer-historian James Willard Schultz offers further evidence that such indigenous geography concerns were long-standing issues for the Blackfeet leadership, and that they were intent on actively sustaining and producing Blackfeet geographies. In *Signposts of Adventure: Glacier Park as the Indians Know It*, Schultz (called Apikuni, or Spotted Robe, by the Blackfeet) indicates that every summer for a decade his Blackfeet hunting partners and friends urged him to take up "this important work," which he repeatedly put off to attend "other and pressing matters" (Schultz 1926). When he finally turned his attention toward their request, they appointed Schultz and a Blackfeet team to work with knowledgeable tribal elders and the Kootenai tribe to "give names" where they had been "neglected" and to "erase all the white names" where they had been imposed (Schultz 1926, 6–7). Against and "amid a rising tide of protest" from Whites not wishing to see such restorations or changes (especially when their own personal commemorations were at risk), Schultz spent months compiling traditional names and their origin stories with the help of his tribal team, including Guardipee's central figure in *Mountain Chief, Blackfeet War Leader*; Mountain Chief (Schultz 1926, 11).

The resulting publication details origins and meanings of 161 Blackfeet place-names inside Glacier National Park, as well as 153 briefly annotated "Kutenai" names gathered for the western side of the park (Schultz 1926, 7). As Schultz's subtitle suggests ("as the Indians *Know* It"; emphasis added), Blackfeet geographical knowledge and spatial relationships were ongoing and not a matter of historic memory or nostalgia. Their concern over place-names encompassed more than just disagreements over geographic labeling. They were expressly worried about the destruction and reconstruction of a world beyond or greater than even their own cultural, economic, and political transformations. Yes, the Blackfeet way of life and relation to the land was threatened. From within their particular frame of spatiality, however, the mountains themselves were also in danger of being transformed.

This brief example highlights the nature of the battles being waged both on and off the reservation lands. Almost immediately following the already power-laden and acrimonious concession of tribal lands that would become Glacier National Park, park officials and capital interests conspired to expand the park boundaries by cutting even farther into the Blackfeet reservation. By 1935, historian Mark Spence tells us, "relations between the Blackfeet and the National Park Service had reached an impasse that remains unresolved to this day" (Spence 1999). During the moments represented by Guardipee's artifacts, "tribal leaders embarked on an unofficial policy of noncooperation," which was soon bolstered by the rise of John Collier's Bureau of Indian Affairs, which supported selected kinds of tribal self-determination. In the case of the Blackfeet, policies of self-determination further empowered conceptions of their tribal geographies, which was already partly fueled by "negative opinions of the park service [and would] become a central aspect of tribal policy and a fundamental expression of Blackfeet national identity" (Spence 1999, 98).

Tensions over the Blackfeet expressing their land-use rights by hunting in the park only lightened when decades of foolish ecological management policies gave way to more holistic understandings. Fire suppression, predator-reduction campaigns, and protection of game animals eventually gave way to the more sound ecosystem-conscious principles that value the role of predators, for example, as well as the importance of diversity and balance of waterways, animal populations, and flora. By the 1950s some park leaders even reversed the content of their blame—if not the blame itself—when they now "complained of Blackfeet hunters not killing *enough* animals" to lessen the impact of overgrazing (Spence 1999, 99; emphasis added).

These tensions around competing spatialities continued in the form of challenges to the "terminal lease" of park lands starting in the 1950s. In the 1990s, Vicky Santana and George Kicking Woman, keeper of the important Beaver Bundle, confirmed a long-standing Blackfeet position that the lands had been stolen (interviewed and quoted in Keller and Turek 1998, 63). In 1991, more than a century after the creation of the park, tribal chairman Earl Old Person likewise explained that the Blackfeet "only sold them the rocks [mountains]" signaling the tribe's sense of retaining full usage, and thus a continued Blackfeet geography within the park (quoted in Farr 2009, 43).

These recurring legal and cultural tensions demonstrate a pattern of contention over two overlapping geographies stretching over a century of resistance. Such historic and cultural frames situate and reveal the spatial significance in Guardipee's use of contemporary maps and the flat, ledger style of visual narrative. The added layer placed over the now-background maps draws attention to mapping within colonial and settler colonial practices, the role of history and discourse, and to overlapping and contested spatial productions. More importantly, the Blackfeet-oriented overlay offers more than just an anachronistic bricolage featuring nostalgic images of traditional culture irreconcilably imposed upon modern cartographies. Given Mountain Chief's direct involvement in challenging the American place-names and working to protect Blackfeet geographies, he is an especially provocative and spatially significant choice for the artist. Guardipee's references to Blackfeet culture are statements of contemporary cultural identities, epistemologies, and ontologies, and thus active claims on or articulations of space. As he states, "Even though there are all these changes on the maps, like highways and the evolution of my homeland, my art tells people that we're still here" (Fauntleroy 2011, 33).

Beyond simple presence, however, Guardipee insists that the Blackfeet "still have our culture and belief systems intact," which is why the use of ledger style and nineteenth-century scenes/figures does not trap his collages in the ever-popular Plains Indian past (Fauntleroy 2011, 33). Asserting that the Blackfeet are "still here" clearly invokes current cultural and physical presence, but also articulates a specific geographic existence, a comprehensive and functional spatiality. And, even as the landscape has evolved, being "still *here*" is a precise Blackfeet locator. The homeland still stands. The geographic precision of their origin and creation stories continue to anchor Blackfeet existence in space. Despite the best efforts aimed at colonization and dispossession, Blackfeet space has not been eliminated and thus must overlap, mostly uneasily it seems, with the constructed geographies of Montana.

5 The Space of Native Art

Greasy Grass, Montana, June 25, 2013

It was a day of speeches and ceremony. Talk of fighting, pain, and land once again filled this space. Dr. Leo Killsback, a Northern Cheyenne, and part of the design team for the Indian Memorial at the Battlefield of Little Bighorn site, told the gathered audience that the Native resistance "was about protecting our way of life and, most importantly, our homeland" (Olp 2014). William C. Hair, a Northern Arapaho representative, argued that "this is the closest we'll ever come to acknowledgement from the government of the atrocities we have suffered" (Bertolini and Ore 2012, 11). As they spoke, three bronze warriors rode off to battle.

The *Spirit Warriors* sculpture stands directly across from the memorial to fallen US Army soldiers, near Crow Agency, Montana, headquarters of the Apsaalooké (Crow) Nation. It stands as both part and counterpart to the Battlefield at the Little Bighorn National Monument. The Battlefield Memorial was once a singularly focused tribute to Lieutenant Colonel George Armstrong Custer and his ill-fated 7th Cavalry. The Indian Memorial, added to this federal historic site in 2003, reframed the commemoration. The thirty-five-by-twelve-foot bronze *Spirit Warriors* runs along the northern edge of the Indian Memorial, which is composed of a semi-enclosed circular mound. Inside the mound one finds names and images etched into dark granite walls merging earth and stone. Except for the bronze sculpture, the Indian Memorial perhaps most closely resembles the aesthetics of the national Vietnam Veterans Memorial. The Indian Memorial, however, stands as the nation's only federal recognition of tribal warriors.

Oglala artist Colleen Cutshall designed *Spirit Warriors* as a striking metal armature that outlines seven figures (see fig. 5.1). In this work, her first in metal, she skillfully renders a flattened yet still three-dimensional version of ledger art techniques. Local journalist Heidi Gease marvels at the open "line drawing" feel, as Cutshall defies her metal medium; Gease calls the piece "the antithesis of bronze," which is generally heavy and solid (2003). The openness of her creation seems appropriate, given its placement among the winds and swaying grass of the plains. Together with the larger Indian Memorial, the sculpture is in a parallel way charged with opening relations: facilitating dialogue about nations and narratives, and about how we define and makes claims on history.

I am drawn to *Spirit Warriors* because I see it as operating via space as much as history. Memorials are often created to articulate a special place, to reflect the idea that a certain location is notable and meaningful. Sometimes the site is important, as in this case. Other times the site is made (more) important by the introduction of the memorial, such as the national Vietnam Memorial. By these placements of articulation, meaning is both created and reflected, and, sometimes, contested. This is true even as the meaning(s) are constantly shifting. The battlefield site itself offers a perfect example, given the fluidity of its name and its competing interpretations. Cutshall certainly captures the notion of fluidity by presenting a moment of beginning (a battle) within the larger tension between contested spatialities. Breathtakingly framed by prairie horizon and the wide sky, the "spirit warriors" rush to secure an indigenous world. The battle being referenced, like many during this era, centered on tribal refusals to return to the reservation. Cutshall's work remembers how American designs on the land conflicted with and were incompatible with indigenous spatialities—something the original memorial explicitly excluded in favor of cleanly nationalistic and racialized memory-making.

The American efforts to confine Native peoples to reservations and destroy bison herds were clear anti-Native geographic projects aimed at remaking the land both materially and discursively (as chapter 3 describes). The attack on Native cultures was also a form of spatial violence explicitly outlining when and where such ways of life could be practiced—generally, in the past, and not within the presumed territory of the United States. Cutshall intentionally captures some of the struggle over conflicting cultures and spatialities by drawing subtle attention to the intimate impacts of the anti-Indian campaigns. She notes that "for any war memorial, you need to have some sense of the humanity that's involved . . . and I don't think you do that with just three warriors riding through the sky" (quoted in Gease 2003). The original call for artist submissions requested just three warrior figures. Cutshall adapted her proposal, adding a woman figure to allude to the families, relationships, and ways of life supporting and motivating the warriors. She still presents viewers three male warriors, mid-departure with their three horses, running off toward what will be a historic, if "last," military victory.

Cutshall presents a lone woman tailing the war party. This figure provides visible support to the resistance effort, handing a war shield to and exchanging possible final glances with the trailing warrior. The depiction of this exchange suggests that the warrior's returning look is not just the practical act of securing the shield, but of affirming the connection to his family and people, a moment of summoning courage by remembering why he is heading off to fight. The mounted/mounting warriors (representing the Arapaho, Cheyenne, and Lakota) head east to confront and ultimately defeat Custer's column and his Native Scouts (Apsáalooke and Arikara). Cutshall reminds us that Native women, although not necessarily part of the fighting, "were present at the battle, providing food, fresh horses, new weapons and encouragement. They assisted men injured in battle and later helped mutilate corpses, which was a ritual believed to kill and block the human spirit in its journey to the afterlife" (Gease 2003). Cheyenne participant and witness Kate Bighead conveyed the story that women

> pushed the point of a sewing awl into each of [Custer's] ears, into his head. This was done to improve his hearing, as it seemed he had not heard what our chiefs in the South had said when he smoked the pipe with them. They told him then that if ever afterward he should break that peace promise and should fight the Cheyennes, the Everywhere Spirit surely would cause him to be killed. (Bighead 2004, 376–377)

Bighead's account informs Cutshall's creation and indicates participation in a larger spiritual realm that is part of the difference between the Native and settler spatialities at stake in this conflict. As the description suggests, the mutilation of Custer's body was seen as an ongoing contestation with him. After his death, he was still subject to the forces that regulated life and death on the plains, and their warning to improve his hearing held ongoing consequences subject to the retribution of the Everywhere Spirit who enforced the laws the Cheyenne had long ago learned to follow. The power of traditional "treaties" between human and other-than-human partners held utmost power. Those agreements and responsibilities were rooted in the land itself, beyond the purview of humans alone, and the resistant Plains warriors and societies expected and hoped to protect a future that sustained these Native spatialities.

Rather than focusing narrowly on the acts of warfare and the glory (and horror) of battle and victory, however, the Native woman figure reinserts the entirety of the tribal world. More than any other element of the piece, local journalist Heidi Gease suggests, the woman in *Spirit Warriors* "represents the tribes affected by this and other battles" (2003). By recovering this tribal entirety, the "humanity involved" centers around Native peoples and cultures oriented toward particular kinds of experiences with and relationships to their lands.

Cutshall's contribution, however, also helps anchor a connecting point between past and present, between indigenous and settler societies. The larger Indian Memorial features a "spirit gate," arranged to open sightline and pathway, intentionally created by the design team to allow transit for the dead. Moreover, *Spirit Warriors* aligns with the gate to provide the spirits of both the Native warriors and American soldiers a direct opportunity for connection, perhaps even reconciliation. The act of generating this kind of memorializing presents an opportunity much like we see in Paige Raibmon's description of the Robert Gray–descended family's remarkable ceremonial apology to the Opitsaht village of Vancouver Island (mentioned in chapter 3). Yet it might more closely parallel the recurring and as yet not fully realized moment for reconsideration found in Satanta, Kansas. As the memorial sustains a portal between earthly and spirit worlds, the paired Indian and cavalry memorials make intersections between coexisting geographies visible and material. The physical intersection reveals spaces held in tension, a "stalemate" that often proves uncomfortable for those accustomed to Eurocentric and tailored narrations.

The memorial marks out a place where current tribal peoples can concretely engage with both the past and the present of that place, with the importance of this battle and of following one's responsibilities to homelands. Now, "there's something there," Northern Cheyenne tribal member Tim Lame Woman points out (Bohrer 2003). He once helped install an "unauthorized" memorial with other American Indian Movement members in 1988. These unofficial memorializing acts reflected long-standing desires to rescript but also reflect how that event and that landscape is understood by many Native peoples. At the monument's dedication, Lame Woman noted "we finally have something, a place for our children to go and see, and it's long overdue" (Bohrer 2003). Public history researchers Jim Bertolini and Janet Ore note, in their application to register the Indian Memorial addition as a national historic site, that the idea for the Indian Memorial at Little Bighorn Battlefield first arose in the 1920s. Mrs. Thomas Beaverheart wrote to then Superintendent [of the battlefield cemetery, Eugene] Wessinger requesting a monument to her fallen father, Lame White Man. The War Department ignored Mrs. Beaverheart's request, a policy to which the National Park Service conformed until the mid-1980s when popular support forced a change in administrative position. In the 1960s, when the National Congress for American Indians (NCAI) and the more radical American Indian Movement (AIM) demanded additional autonomy and enfranchisement for American Indians through demonstrations and lobbying, they raised the issue of a memorial to American Indians at Little Bighorn Battlefield. (Bertolini and Ore 2012, 7)

The tensions worked out in sites like the Little Bighorn Battlefield National Monument rearrange and reconstruct the meaning of this event and its site. The space itself shifts dialectically to reflect and reshape the identities of those giving it meaning.

The Native victory over Custer in 1876 was an unexpected loss at a time when the frontier was rapidly closing and only a handful of Native peoples were still defiantly surviving in the northern plains and southwestern deserts. This battle's loss marred the nation's otherwise ebullient centennial. National discourse and military redoubling reflected the embarrassment and symbolism of such a defeat. The loss rendered a centuries-old colonial project incomplete. It rendered the nation's geographic project and its inevitability suspect. It momentarily challenged the "self-apparent" superiority of Western culture and European Americans and thus led to renewed pursuit of this presumed natural hierarchy. The deferred colonial project and those paused notions of historical trajectory were disruptions of White racialized and national space, and uncomfortable affirmations of Native space. While military defeats can be reconciled as part of the cost of war, this concrete proof that supposedly primitive tribal space could persist and repel the certainty of a predestined American geography was unacceptable.

We can easily trace the instructive overlap and changing of names at this site as a way of observing its spatial constructions. As a focal narrative point for the US nation-state's privileged historical centrality, it was quickly tabbed as Last Stand Hill and physically marked by a large white monument covering the relocated graves of army soldiers. In 1946 it was named the Custer Battlefield National Monument. The name was changed again in 1991 to the Little Bighorn Battlefield

National Monument, in recognition of an indigenous perspective that viewed the battle through a decidedly different lens. The Lakota, of course, have long referred to the battle as happening at Greasy Grass—marking the event using a preexisting Native cartography. By 1999, grave markers were installed for two of the fallen Native warriors, Lame White Man and Noisy Walking. A few years later, ten were marked. When the 2003 Indian Memorial was completed, 127 years after the battle, the list of all seventy-five or so fallen Native warriors (and long ignored oral histories and pictographs of the battle) were installed.

Whether Custer or his final battle were actually strategically important matters less than the fact that it has taken on such cultural significance and sustains its place in historical narratives and popular lore (Flores 2009). Americans scripted the battle at Little Bighorn as a thoroughly nationalistic event that no one survived, generating defeat sympathy and conveniently ignoring the thousands of Native people who came away alive and victorious and telling stories. For a time, historians ludicrously even positioned "Comanche," a solitary army horse escapee who returned to Fort Riley in Kansas, as representative of the American experience. The horse was treated with historic reverence, eventually being "stuffed and put on display at the University of Kansas" upon his death (Brooke 1997). In short, the man (Custer) and also the battle itself have both been treated in a variety of ways, from reverence to ridicule. Obviously, Custer remains a shared figure for non-Native and Native alike. Naming the site after Custer, however, illustrated a specific kind of meaning-making. Stepping back, we can better see how the treatment of Custer, the horse Comanche, and the battlefield reveals a struggle between American and indigenous spatialities.

More than just a temporal point of transformation, this event and the Little Bighorn Battlefield serve as case studies on the tension of overlapping spatialities. The possession of that space and its meaning are interrelated. It indicates a sense of ownership and announces the spatiality being privileged. Since the victory at Greasy Grass, this space has been filled with conflicting and emotionally laden sets of meanings. For the first several decades after the battle, the dominant narratives commemorated this space as a massacre site. Most understand that the battle at Little Bighorn, and all Native-US wars, were at least partially concerned with land and resources, although this typically gets subsumed to the narratives of cultural clashes and the inevitable Native submission to "civilization." These narratives and the encounters being narrated are, of course, only materially meaningful as spatial acts. Nevertheless, the discursive and historical trajectories tend to emphasize the abstractions of an ontology of global social evolution with a universalized teleology (in short, Manifest Destiny as being neutral, and good for everyone).

For Native people concerned with this site, the physical space and their spatialities needed reconciliation. The lack of resolution among spiritual, cultural, and material resulted in "illness." As Cutshall asserts, "Native people have needed a place there to connect to, and you just couldn't with that Cavalry orientation. You just got sick" (Gease 2003). The dual or newly synthesized memorial(s), by occupying the same land, now more easily forces the question of geography and spatiality. The different meanings and relationships to this space are laid bare and given the simultaneous presence already in place "there." Rather than simply trying to answer questions about meaning with authority and finality, the site is now explicitly involved in fluid and contested inquiries of meaning-making and spatiality. What is this place? What happened here? Why is it important? Whose space it this? Who belongs here? Who are we in relation to the space? Who decides how it is defined? What are the implications of our various answers?

Even with the contestations and changes, Little Bighorn has just one access point for the United States: a historically contained temporal and cultural reference. Without this battle, this site is not part of American discourse and history. It remains, in effect, Greasy Grass Creek for a good while longer. The presence and death of Custer in this place, however, marked that moment and that space in a new way. Now it has been marked again, led by attention to an indigenous geography that highlights the tension of histories and spatial practices.

Why Installations?

Following the model suggested by Colleen Cutshall's amazing work, I want to use the rest of this chapter to present the work of two additional Native artists using public installation art to disrupt settler spatialities and to mark and reestablish indigenous geographies. Cutshall, Edgar Heap of Birds, and Bob Haozous each offer confrontational creations that physically illuminate how a space is being produced and draw attention to the co-constitutive relationship between space and identity. Using public installations, these artists contest and reshape dominant spatialities, both implicitly and explicitly recovering tensions embedded within indigenous and settler colonial constructions of space. Much like the artists working with maps (discussed in chapter 4), their work draws attention to how spaces exist as overlapping and contested realities and suggests the need for greater consideration of how geography is a vital category of analysis toward making sense of such tension.

In terms of the practical methods the artists use to force this "conversation," we should note the most salient distinctions between the artists in this chapter and those in the previous chapter. First, we see a transition from the use of maps to the use of public installation. This move is a shift in genre and media, but also a shift in the explicit and hands-on engagement with spatiality, for both the artists and viewers. A map is a representation of spatiality in two dimensions. Its encounter requires a conceptual bridge from representation to the material world. A public installation piece is a three-dimensional map that, because of its dimensional and locational advantages, can work as a representation of spatiality while physically occupying a site-specific space. In brief, maps tend to work from a larger conceptual scale toward the smaller, while installations start from smaller scale and work toward applicability on the larger scale (beyond the immediate site). Installation pieces, then, serve as spatial markers themselves, and thereby tend to resemble the street signs discussed in chapters 1 and 2 as well as the map creations discussed in chapter 4.

I start with Edgar Heap of Birds, whose work creates conversation between the kinds of work being done by the Native and non-Native street signs outlined in those earlier chapters. According to art scholars Nicolas De Oliveira, Nicola Oxley, Michael Petry, and Michael Archer, installation art "rejects concentration on one object in favour of a consideration of the relationships between things and their contexts" (De Oliveira et al. 1994, 8). One of the core ways that installation art concentrates on relationships and contexts is to deemphasize art hanging on the museum wall. Installation artists typically seek to shape an experience so that the viewers become consciously implicated in the artistic process. Often they locate such work outside of the standard settings—galleries, museums, studios—in order to draw attention to the relationship between viewers and the world they interact with, and remind them of how placement and engagement shapes the encounter. Installations strategically become part of our spaces in order to help us recognize the contours of that space and ways of embodying our spatialities, and then to challenge them and their assumed authenticity.

These opening points of discussion lead us nicely toward the work of Hachivi (sometimes written as Hock E Eye V) Edgar Heap of Birds, a Cheyenne/ Arapaho artist who is currently a professor of art and Native studies at Oklahoma

University. Heap of Birds has become notable for producing public installations consciously sited to simultaneously raise localized and more national- (sometimes global-) scale questions about colonialism, dispossession, discourse, indigeneity, racism, and visuality. His works typically aim to stimulate discussions about collective identities and narratives. I am particularly interested in his work with public signs, which range from street and parking signs to billboards. His intentional appropriation of public and street signs—those most dominant tools for everyday spatial ordering—interrogates what some call the "taken-for-granted infrastructure of daily life" (Rose-Redwood 2009, 461). In brief, he hijacks everyday tools that "invisibly" shape our world, works to make them newly visible, and then turns them upon themselves.

Installing Recognition

In contrast to the creation of the Indian memorial at the battlefield at Little Bighorn, Heap of Birds focuses his work on the collective forgetting that happens via everyday practices and in everyday places. Memorials operate as physical interventions in space to combat memory "loss," and to continually embed specific meaning where it constantly seeks to be changed and lost. Memorials try to stabilize space, to normalize and materialize certain spatialities. In contrast, Heaps of Birds facilitates a destabilization of dominant cultural geographies via everyday spatial markers such as street signs. He works to show how the same cultural and spatial processes that produce memorials are in operation in these mundane technologies as well. In noticing these mechanisms, he admits to being "fascinated by how gullible we are, to believe the propaganda" and that "whatever you have in print, people believe it very readily" (Heap of Birds 1999).

Taking advantage of this official reliance on text-as-truth ("when you're being subversive, that's the best arena to use"), he works to harness the very same power to craft spatial meaning via unexpected reinventions. In these ways, Heap of Birds' public installations overlap with Cutshall's officially sanctioned memorial, but perhaps more closely align with those unofficial efforts by American Indian Movement members who temporarily marked Native warriors' grave sites and intentionally sought to disrupt the unchecked memorialization of Custer. Whereas the American Indian Movement memorials were removed because they were deemed unauthorized, Heap of Birds' works are solicited and commissioned. Perhaps because they are understood as art, and temporary spatial insertions, they find a welcome place in the interruption of landscapes.

W. Jackson Rushing III notes that many of Heap of Birds' early public installation and concept art pieces actively engage in a kind of symbolic "reclamation
of social space" (Rushing 1999, 375). Janet Berlo and Ruth Phillips suggest that
Heap of Birds holds a deep appreciation for the long and important history
of Native artists using modern art forms to convey "insurgent messages." They
quote, for instance, Heap of Birds' insistence that imprisoned Cheyenne ledger
book artists used their drawing as a way of "defending native peoples," rather
than just documenting their incarceration experiences or reminiscing over previous times (Berlo and Phillips 1998, 214). As Heap of Birds has stated, Native
artists often "find it effective to challenge the white man through use of the mass
media [and that] the insurgent messages within these forms must serve as our
present-day combative tactics" (Heap of Birds 1987, 171). By redesigning ordi-
nary streets signs and posting public "orders," Heap of Birds boldly calls out the
role that uneven social relations continue to play in the suppression of Native
cultures, peoples, and geographies. He reminds or points out to viewers where
they are located, both physically and culturally, in relation to indigeneity and
settler colonialism. His art renames public spaces, marks their ongoing production as settler spatialities, and forces an acknowledgment of persistent Native
geographies.

Heap of Birds' signs mimic authoritative street signs in presenting simple,
direct statements with no artistic elaboration that might distract from or contextualize the message, or conceal the co-productive relationship between sign and
viewer. He places his viewers in a position to contend with the inconsistency of
an authoritative sign displaying an anticolonial message, for example, a Toronto
billboard emblazoned with "Imperial/Canada/Share/Stolen/Lands" (Heap of
Birds 2009, 32). In the context of reminding people of Native geography, Heap
of Birds seeks to "[re-label] the landscape to exile the white viewer" (Ohnesorge
2008, 59). In online statements for the Walker Art Museum in Minneapolis,
Heap of Birds explains that his public creations are specifically intended to generate discussion and disrupt simplistic, and especially colonial, narratives. He
acknowledges that "public discourse is part of the work" and actively works to
generate an awareness of the relationships among artist, art, audience, and context (Heap of Birds 2007).

Heap of Birds understands that producing truly engaging public art requires
fostering intersections. The challenge of creating such opportunities drives his
artistic process and shapes his final creations: "I expect it, I deliver it, and we
deal with it. It's not just the work, it's what happens between the art and the

public—that's public art" (Heap of Birds 2007). Such engagement requires a careful balance of presentation and intrigue. "If you're too explicit," he argues, "people just turn the page and go to have lunch. They don't really dig to find anything. I think you have to hit that edge, where it, just visually, makes you wonder 'what's that? Should I look at that?' and then investigate what that means" (Heap of Birds 1999).

I want to first take a look at one of his more standard non-installation pieces, *American Leagues* (1996). This painting nicely illustrates Heap of Birds' willingness to address controversial topics, and to do so in a relatively straightforward manner. As a critique of Indian mascots, Heap of Birds re-presents an intentionally grotesque version of the Cleveland Indians baseball icon. Long the target of activists concerned with representational force and the implications of Native peoples serving as mascots, the image is placed on a white background embellished by wild marks resembling smudges from bloody hands. The four edges offer minimal text (clockwise, starting from the top): Cleveland Indians/Human Beings/Not Mascots/Value. The phrase "Smile for Racism" boldly announces the artist's position on the use of mascots. In all, he echoes the long ago articulated observation by Vine Deloria, who noted that Native peoples' biggest problem in relating to Whites was White inability to see Native people as human beings (Deloria 1969, 2). They could only see, he notes, "mythical Indians." Heap of Birds painting makes a similarly forward assertion and reveals his contentment with moralistic and didactic approaches to audience engagement when deemed necessary. The weight of Deloria's observation means that Heap of Birds, and all Native artists, must confront the mythologization of Native people and thus constantly confront a serious need to inject Native realities where they are invisibilized and ventriloquized. As Heap of Birds argues,

> even as these grave hardships exist for the living Indian people, a mockery is made of us by reducing our tribal names and images to the level of insulting sports team mascots, brand name automobiles, camping equipment, city and state names, and various other commercial products produced by the dominant culture. This strange and insensitive custom is particularly insulting when one considers the great lack of attention that is given to real Indian concerns. . . . To be overpowered and manipulated . . . [and] become a team mascot is totally unthinkable. (Quoted in Wood 1998, 67)

I draw attention to this piece to illustrate Heap of Birds' direct approach and political tactics, but also to introduce the embedded spatial critiques within such work. *American Leagues* may not be immediately recognized as a spatial project, seeming to be confined to the realms of the cultural or political. Yet given the relationships among culture, politics, and space, this work is highly attuned to a colonial and racialized terrain that shapes the very experience of Native peoples (indeed all peoples) and the ways in which indigeneity figures into dominant constructions of nation, citizenship, Americanness, and the reach and force of discourse. As Lucy Lippard points out about his work, "even when the subject appears to be something else, land is the bottom line" (Lippard 2008, 22–23). This might be expected, since the positionality of Native peoples in the United States and the material impacts of colonialism and racism are directly rooted in the spatial projects that craft the nation. In relation to mascots, this frame further helps explain why "Indians" are needed to secure American identities, how Native peoples are narrated with that history, and how it requires the dismissal of tribal territory and sovereignty toward models of multicultural citizenship. Mascots are symbolic tools for narrating the consolidation of a national landscape and the exultation of a supposedly equalized multiracial citizenry.

Compared with *American Leagues*, much of Heap of Birds' work tackles spatiality and its relationship to indigeneity in a much more explicit manner. His creations tend to reflect an engagement via the medium itself, in that they fully embrace the form of installation rather than museum art. Consider one of his early set of works, the "Your Host" series. Starting in 1988, he installed a series of signs in New York's City Hall Park. Heap of Birds borrows discursive authority from public signage to re-announce Native spaces under the very feet of and everywhere surrounding onlookers, visitors, and passersby. This series has now been extended for three decades in places across the United States and around the globe, with the latest being installed in Anchorage. As in the original New York installation and every site-specific installation since, the Anchorage signs reference the indigenous peoples of what is now a US state: Haida, Tsimshian, and Unangax̂ (Slocum 2007). We first see the place-name "Alaska" presented backward, intentionally causing pause and rendering the familiar strange, perhaps undoing this name as we read it backward. This attention-catching callout or hailing of the public is followed by a simple five-word phrase: "Today Your Host/Is/Unangax̂."

The signs remind residents that they are being "hosted" by the respective local indigenous group. Their messages are articulated in a standardized, almost unnoticeable "wet grass" park sign format (see fig. 5.3). There is no further explanation, certainly no apologies, and no explicit direction on what to do with such information. The work expects, perhaps, that the reader simply absorb and understand the information provided. The intent, of course, is that such information will initiate the kind of courtesy and deference that any guest should show toward a host. Hosting practiced here and now. Which begs the question for the viewer: Is *guesting* being practiced? Regardless of reception, Heap of Birds' signs proclaim in dramatic fashion the unequivocal persistence of indigenous geographies in a place as culturally complex, historically layered, and materially dense as New York City, as well as in a place undeniably marked with Native presence like Alaska. They also implicate everyone who can read the signs in now deciding how they will re-navigate this re-grounding.

While the simplicity of his "Your Host" series does not extensively interrogate the complex and layered histories and their spatial implications, it does call for recognition and some form of contemporary engagement. What does one do when confronted with indigenous presence precisely where it is least expected, in urban centers and in the most successful regions of removal and colonial settlement? The fact that Heap of Birds includes the word "Today" in his selectively sparse textual creations elegantly indicates its importance to the overall message and intended effect on the viewer. He is firmly pointing to the settler colonial present, because settler spatialities are ongoing, as are indigenous ones. His method of creating such pieces reflects this framing as he intentionally works with local tribal peoples to responsibly incorporate their cultures, lands, languages, and peoples as part of his interventions (Blomley and Heap of Birds 2004, 800). He has often referred to his approach as an explicit effort to "commemorate or honor nations" (Heap of Birds 1999). "The first step," he tells us, "is to bring the indigenous presence back to lands and urban sites which were lost to the white invasion" (Blomley and Heap of Birds 2004, 800). Clearly, he is drawing attention to an occupation, now. Today.

Rather than accepting that his work is somehow objectively politically charged, Heap of Birds reframes settler geographies as being the charged sites of colonization and indigenous displacement that require overdue investigation. Thus, "cities are locations of the sign pieces because often the sites within cities have high value because of the colonial power's wealth. They are charged locations that can be implicated in an unsavory history of conquest" (Blomley and Heap of Birds 2004, 800). In *Reclaim* (1996), for example, he borrows aesthetic force from standard green and white highway mileage signs by placing his appropriated sign along the roadside (see fig. 5.4). Under commission of the

Neuberger Museum in Purchase, New York, the sign was placed at the main entry of Purchase University grounds, making it available twenty-four hours a day at the intersection of Lincoln Avenue and the West Road portion of the main campus loop. In contrast to the "wet grass" sign, the text of this piece more explicitly interrogates indigenous geographies and boldly suggests decolonized spatial futurities.

On the one hand Heap of Birds asks about the spatial fate of New York. Purchased? Stolen? While the piece is ostensibly solely about the geography of what becomes New York, it actually expects the viewer to locate themselves (individually, collectively) in relation to the land. For the Biennial Exhibition at the Neuberger, Heap of Birds argues that his mock interstate sign marks "vehicular movements over colonized lands, methods of territorial procurement and the spiritual reinstatement of rightful indigenous awareness within the State of New York" (Neuberger Museum 1997, 8). In short, it pushes us to note our assigned meanings or assessments of this space. In placing question marks in the usual mileage position, he provokes an open-ended response not only to the question of history, but also to the means by which we can measure when we might (or might not) arrive at our destination.

The final indicator, "Reclaimed?," in particular, hints at the ongoing refusal of Native peoples to simply concede the loss of their lands. Consider the exceeding relevance of persistent Iroquois claims to sovereignty, most recently publicized via their refusal to attend an international lacrosse tournament when England would not recognize their passports (Kaplan 2010). Heap of Birds offers us brief insight into the continued understanding of Iroquois geographies, those being expressed and acted on, and those being kept in the sovereign imagination awaiting future manifestation. Reclamation already practiced, and a possible future reclamation of additional material consequence. In essence, Heap of Birds merely pulls back a colonial veil that obscures what is already practiced, even if only within a cultural framework of indigenous communities largely denied by others. As he acknowledges, he is just "there to translate it" (Heap of Birds 1999).

Heap of Birds' art explicitly works to both assert Native geographies and to uncover the role of mainstream spatial markers in producing and maintaining a particular, racialized, and sanitized version of colonization. He re-marks what seem to be public signs normally considered little more than civic infrastructure and navigational aids, subverts the standard messages, and reveals standard signs as works of popular culture and settler functionality. He readily recognizes their constructive power and therefore the appropriative value of such mechanisms, since public "signs are all thought to be true" by the average citizen (Rushing III 2005, 376). Like Jaune Quick-to-See Smith (discussed in the previous chapter), Heap of Birds crafts an engagement that almost requires viewers to pause and reflect on what they are seeing. Those moments of pause produce, at the very least, inquiry about the message and the intended relationship between the viewer and sign.

No Coffee Cups and T-Shirts

Bob Haozous (Chiricahua Nde'/Apache, Diné, Spanish, and English) offers us a variation of memorials like that produced by Cutshall, and a different approach to public installation work addressing indigenous space. Haozous is a highly regarded sculptor and public installation artist from a family of artists, most prominently including his late father and fellow sculptor Allan Houser. He draws from a range of materials, although I focus on just one of his metalwork sculptures that broadly represents his signature medium and style. As with Heap of Birds and Quick-to-See Smith, many of Haozous's pieces reference familiar objects reworked to beg greater depth of engagement with its viewer. Whereas Heap of Birds directly appropriates authoritative directives in the form of street and public information signs, in the work I address here, Haozous references both memorials and the aesthetic and property-claiming function of spatial markers like southwestern ranch gates.

Haozous produces what might be called monuments, although his large-scale sculptures are more focused on raising hard questions than simple "remembering." This flies in the face of most memorials, which work to arrest meaning, even while his pieces do actively deploy the tactics of memorialization and memory-making.

Gate/Negate (2000) is currently on loan with the Santa Fe Capitol Art Foundation and rests on the northeast capitol grounds. This bronze piece is eighteen feet tall and composed of two main elements: a solid trapezoidal base and a rectangular frame or "gate" partly filled with a number of smaller, individual cutout shapes. The gate frame itself is sparse and largely utilitarian, save the chain rings fastened to the top corners. Haozous describes his general use of rings or chains as a "symbol of slavery or oppression" provided to offer viewers "handholds, or clues into the [political nature of the] piece" (Anon. 1990). The cutout shapes within the gate's outlining frame provide its main visual complexity; dollar signs and crosses interconnect a dozen negative-space human profile silhouette plates. The silhouette plates are also embellished by cutout airplanes flying sky-ward, flattened stars, and bulging spheres (that he identifies as bullet holes). As another clue to the political nature of the work, a segment of razor wire tops the gate. The trapezoidal base (described below) is constructed out of corrugated sheets with writing hand painted across all four of the visible surfaces.

In all, *Gate/Negate* is positioned and shaped like a memorial or monument, although Dean Rader notes that the staff of the art collection, while uncertain of the piece's meaning, assured him "that it was not a 'monument'" (Rader 2011, 199). Much like the Indian Memorial at the Little Bighorn Battlefield, however, this would-be memorial counterpoints nationalistic remembrances that depend on the logic of colonialism. Consider, briefly, Haozous's insistence on placing razor wire atop *Cultural Crossroads of the Americas* (1996), among other of his works. Razor wire sometimes makes all the difference, it seems, between recep-tivity and trouble-making. In *Cultural Crossroads*, Haozous took a page out of Heap of Birds' book, constructing a twenty-six-by-twenty-nine-foot billboard for public engagement in conjunction with the University of New Mexico and the city-funded Art in Public Places program. The powerful iron cutout swiftly tackles the US-Mexico border, environmental degradation, capitalism, global-ization, indigeneity, and culture. The inclusion of razor wire, however, became the breaking point for some university representatives, who saw the addition as a symbol of divisiveness and conflict rather than the image of multiculturalism and connection they desired (Mithlo 1998, 59).

The ensuing conflict over political interpretations and artistic integrity led to a nasty court battle, after which Haozous was forced to remove the razor wire or risk being denied nearly half his commission. Native studies and art scholar Nancy Mithlo, Haozous' spouse and sometime chronicler, quotes Haozous as saying that it is "absolutely essential" for a Native artist "to remain honest—either that or make coffee cups and T-shirts" (Mithlo 1998, 60). Reluctantly conceding in this case, he argued that his detractors "just want to make art digestible to the tourist crowd" (Willdorf 2000). Mithlo points out that while razor wire was indeed used with the imprisonment of Nde' peoples, it was also used by the Nde' for their cattle ranching (Mithlo 1998, 59). She rightly suggests that the meaning of this material is not easily fixed. It is certainly true that determining what and when objects represent divisiveness is a subjective assessment and a frequent tool of political grandstanding, especially when considering artistic creations.

Given Haozous's intentional use of art as cultural critique, however, I do lean toward reading its inclusion as intentionally provocative, which is still neither equivalent to divisive nor justification for censure. Haozous has made it clear that in pieces like *Gate/Negate* he uses "a coil of razor wire symbolizing our [US] isolationism," which is another way of noticing the policing of space and "appropriate" spatialities (Haozous 2007). The political edge to Haozous's work represents its most compelling element. Without this, it fails to be coherent. So, in the case of *Cultural Crossroads*, I take Haozous's word that he sees and uses symbols of barriers, what he calls "antiquated obstacles," in order to initiate critique (Haozous 2000). Illustrating the most immediate, exclusionary function of the gate, Haozous, particularly through the use of razor wire, draws attention to structures of separation as well as to the racialized and settler project of proclaiming land rights. The razor clearly symbolizes the intertwined sense of both spatial control and violence.

Gates can be and are used simultaneously as points of denial and as points of invitation and welcoming. In *Gate/Negate*, we see a tension of meaning present in the penetrable iron gate (or window/picture frame) sitting atop the impassible base. The gate presents the passage or site of entry for immigrants and settlers who have come to occupy the US American territories and other indigenous lands. Thus, the gate is a doorway, the place for entrance. This parallels the "spirit gate" element of the Indian Memorial at the Little Bighorn Battlefield, which also references gates as intentional opportunities for movement rather than just mechanisms designed for exclusion. The fluidity, movement, and spatial accessibility of the gate portion of *Gate/Negate* reflects opportunity for connection as it presents the wholesale transformation and mobility of American space crafted through immigration and settlement.

In this piece, however, Haozous is not necessarily interested in reconciliation or extending welcome in a simple manner. That remains, as-yet at least, one step away. Before connections can be made, the disjuncture must be mapped and mended. As the title asserts, the key concepts in this piece are the gate and the act of negation. As a historical mechanism for controlling access to land and nationhood, and as the razor wire crown suggests, entry is not guaranteed. Access is not always welcomed, or without risk. Thus, the gate is a doorway, but in this case, as the site for exclusion, the threshold site where one may be denied. Further, being permitted initial entrance does not preclude later expulsion. Further still, any shared formality of entrance does not promise equal experience within. This is especially true since different bodies carry "borders" with them that serve as both shorthand and mechanism for spatial orders (Chang 1997).

In short, the gate is a flexible tool of control that shifts according to specific and varying interests or conditions and that can signify a spectrum of differential modes of inclusion and exclusion. Indigenous peoples, as the strangest and most ironic of admissions to the US nation-state, exemplify how difference shapes encounters with the gate in the modern world. Haozous subtly hints at this layered and ongoing complexity in the carefully reconstructed "ethnic" and "sexed" (male and female) portrait silhouette plates. These plates nicely reference racial typology guides based on the "science" of physiognomy once used to assist in distinguishing between groups and accurately assessing (really, assigning) racial character traits, while also mirroring the related eighteenth- to mid-nineteenth-century domestic art of silhouette portraiture commonly referred to as "shadows."

Wrought Irony

In *Gate/Negate*, Haozous simulates a conversation about movement, immigration, and access by insisting on the question of indigeneity via the trapezoidal base of this sculpture. Through the relationship between gate and base, he reminds us of our individual and collective implication in enacting this specific form of negation; the contemporary reproduction and maintenance of colonization and occupation. This dialogue between base and gate also represents larger tensions among the multicultural state, colonization, and indigenous geographies. In all, Haozous crafts a representation of how the so-called founding of democracy and liberty is materially framed by colonialism's logic of elimination, and thus has not resolved settler colonialism's contradiction of indigeneity.

To elaborate this point a bit, I want to first say more about the gate portion and take note of Haozous's use of metal cutout work in relation to southwestern ranch gates. Although he has not, so far as I can determine, made this linkage, he seems to intuitively evoke the form and sentiment of these unique southwestern structures found so prominently in Arizona, Colorado, New Mexico, Oklahoma, and Texas—all places where indigenous and colonial spaces have more explicitly and intimately overlapped in mutually defining fashion for half a millennium.

Haozous masterfully mimics and redirects the unique flat iron cutouts and silhouette style in the way ironwork gates have long decorated and embellished the entryways of expansive ranchlands (see fig. 5.8). Yet ranch gates are specific to non-Native dominated space. As Daniel Olsen and Henk van Assen note, based on their extensive travels through the rural ranchlands of the southwest, "the fewest gates appeared on Native American reservations" (Olsen and van Assen 2009, 23). Let me be clear that Native people have been ranchers and "cowboys" for generations. Whether Native or not, however, the rancher who places a gate at the entrance of an enclosed territory participates in a European American tradition bringing together American, English, Mexican, and Spanish values of private property, economics, and cattle ranging (Olsen and van Assen 2009). The lack of precontact land ownership practices, along with the institutionalized dispossession of Native communities of the most productive and desirable grazing lands, means such gates are a rarity in Native America. "In contrast to the dominant culture," Olsen and van Assen suggest, "Native Americans apparently do not feel the need to put their name big on their land" (Olsen and van Assen 2009, 23). In short, such signs tell us something about how spatiality operates and manifests differently in these cultural and racialized contexts.

Geographer Kenneth Helphand goes on to consider "what is being commemorated" by the ranch gates. He quickly concludes that the southwestern ranch gate is a marker of dispossession itself, noting that for non-Natives "settling the land was itself [seen as] a triumph, accomplished at the expense of the native inhabitants. Settlers built and established a place in the landscape, and made a home and an economic enterprise, for a ranch is both. The passing under an arch/gate celebrates an event, the triumph of the pioneers' arduous and difficult work" (Olsen and van Assen 2009, 84). In a 1929 text, *Wrought Iron in Architecture*, Gerald K. Geerlings argues that decorative ironwork did not take form in the United States until after conquest was secure. After noting the more "simple, practical nature" of most early American ironwork in contrast to its aesthetically complex European counterparts, he states without irony that the "more ornate forms—balconies, fences, gates, grilles—were a development of the late eighteenth century, when the worries of too much Indian and too little corn had been mitigated" (Geerlings 1983, 143). While Geerlings is focused on urban ironwork and its development on the east coast, his observation indicates that the craft and material production of decorative ironwork including ranch gates is dependent on colonization and Native dispossession. Wrought iron art emerges precisely to mark spaces of indigenous elimination. Thus, such work is precluded before Native elimination (however incomplete in actuality). Haozous's (intentional or coincidental) allusion to these metalwork techniques and ranch gate aesthetics as tools in the critique of settler colonialism, then, offers a profound inversion. Like Heap of Birds and Terrance Guardipee (discussed in chapter 4), Haozous reappropriates and deploys colonial tools to draw attention to the process of colonization and then make anticolonial declarations.

Haozous's inversions extend to the identities crafted by and through structures of Western individualism. All ranch gates can be read as repeating and re-marking spaces of colonization and elimination, and are thus clearly productive of spaces and larger social identities. Daniel Olsen observes how iron ranch gates rely on a constructed model of self-identity whereby an individual can say "[I] don't need others to define me" and thus is capable of "determining his or her own context . . . [and] constructing their own identity" (Olsen and van Assen 2009, 25). This is ironic, as Lippard notes, since a great of deal of ranch lands are actually leased out public lands generated by socially, legally, and politically engineered acts of colonialism and managed by governmental oversight and contract. Many reservation lands are likewise leased to ranchers, offering perhaps the most blatantly state-based power-laden spatial arrangement even under trust land status. As a whole, ranching is heavily subsidized and thus deeply situated as a socially managed practice (Olsen and van Assen 2009, 34).

In her introduction to *Ranch Gates of the Southwest*, Lucy Lippard notes that despite the seemingly utilitarian nature of this vernacular ranch gate art, which serves as warning sign against trespass, "the notion of *cultural* trespassing . . . is inherent" (Olsen and van Assen 2009, 33; emphasis original). The inherent trespass indirectly signals that the ranch is anything but the exclusive work of American individualism. Set against the understanding that the ranchers have unproblematically and rightfully become part of the land through their own hard work and collective cultural inheritance, many ranchers actively "[collect] the 'Indian things' found on 'their' lands." The contradiction of collecting indigenous "artifacts" on lands claimed as one's own highlights the tensions Haozous notices when he counters that "the self-made man in a country of immigrants thinks he owns himself, but he is only divorced from his past" (McCloud 1995). I would argue that "he" is divorced from the present as well, given the continued interrelations necessary not just for social standing and legal recognition of title, but also for the continuation of the American individualist prototype.

It is in this light that Helphand situates the ranch as a postbellum institution, and notes that "livestock as capital converted the western grasslands into a commodity" (Olsen and van Assen 2009, 67). He also notes the radical and ironic exchange of one "grazing bovine" (the bison) for another (the cow), which parallels the exchange of one way of life and one group of people for another on this land. Despite the sense that hardworking ranch hands carved out these exchanged realities by sweat and toil alone, numerous social structures and support mechanisms actively helped to produce the large-scale national outcomes and fulfill coherent geographic projects, including the Homestead Act (and its legislative extensions), the establishment of regional and national parks (for example, Yosemite in 1864, Yellowstone in 1872), the ongoing creation of reservations, Indian allotment policies (creating "surplus" lands from 1882–1934), Exclusion and Alien Land laws targeting Asian immigrants and peoples, and the denial of emancipation land promises after the Civil War (Romm 2002). All of this constitutes a thoroughly racialized and restricted process. Together these policies thereby structured and produced a profoundly racialized national geography, depended on a colonial reconfiguration of space, and allowed some individuals access to land and commerce and denied access to others. Even more fundamentally, all such access or restriction depended on Native dispossession.

Silence Is Not Innocence

In *Gate/Negate*, we see a repeat appearance of razor wire as Haozous explicitly works out his frustrations over the court proceedings and the reception of his concepts embedded in *Cultural Crossroads* (Haozous 2007). In yet another politically direct feature of *Gate/Negate*, Haozous presents a base (perhaps tombstone) marked with 460 "extinct" victims of European contact and settlement. He paints the names of Native nations that literally perished under the weight of American colonization. In this way, the base and the entire piece function as a historic atlas of settlement and depopulation on this continent. Recalling their unknown and unspoken extinctions (at least as culturally distinct peoples) serves as the base of this monument, and thereby represents the foundation of the nation. The gate, structured much like a glassless window frame, rests firmly on this negation, a negation that is now actually doubled; the physical and cultural elimination of those peoples replicated by the subsequent, ongoing negation of their name and this history. Even when memorialized here by Haozous, these names will be obscure to most viewers. Without guidance, few will connect the names to indigenous peoples of North America. During his visit, Rader noted that fellow viewers ultimately concluded the names were simply "made up" (Rader 2011, 199). After several days of speaking to visitors about the symbolism of the extinct tribes while he painted the names, Haozous "left with the realization that most viewers have no knowledge of a true American history" and "feel no responsibility for what has and is continually happening to indigenous people world wide" (Haozous 2007).

If that history and the spatial implications are recognized, however, a viewer can discern an important relationship between the gate and negation. Only when supported by such indigenous negation can the gate offer entry/refusal to those multiple ethnic groups who might pass through as immigrants, stretching back to the first European arrivals and extending forward to current immigrants from across the globe. Those immigrant entries can then overwrite the forced entries (enslaved peoples and colonial subjects) that forged the colonial encounter and foreclosed indigenous futures and spaces. Haozous offers a documentation of these historic and spatial processes, of movement and land, of ethnicity and nationhood. He outlines precisely the challenges of those who are interested in social justice and yet fail to simultaneously work to reconcile indigeneity.

Unlike a standard memorial, Haozous's base/tombstone transgresses and remembers those who have been actively and purposefully forgotten, and by doing so refuses those Native eliminations that are extended and repeated through mundane and continual acts of forgetting. He offers a kind of anti-memorial dedicated to remembering what the settler colonial nation must forget to pursue its teleological thrust into a future where Indians/Native peoples are already and have long been rendered absent. In Rader's reading of this same piece, he similarly deduces that *Gate/Negate* effectively "rewrites the present by unerasing the past" (Rader 2011, 200). In a geographic sense, however, Haozous is also remaking the nation by remapping Native land and recentering their denied presence. Given the scale of this sculpture, viewers will come most physically face-to-face with this absence/presence in the form of the painted base before their attention is ultimately drawn upward to the gate. Haozous clearly intends to intervene as he names the tribes, even if he can neither fill the immense and intentional voids of national forgetting nor reclaim indigenous lands.

The late anthropologist Michel-Rolph Trouillot argues that memorials and representations of the past are in fact really about the here and now (Trouillot

1995). They are most fundamentally moments of articulation about ourselves, and thus their meaning is largely determined by what we do today regarding the structures of power and disempowerment and our participation in producing inequalities and committing acts of violence. Otherwise we "renew" them. Haozous suggests this same danger of renewals in his observations of the viewers encountering his sculpture. This stance of understanding history-as-present removes us from pretending to take "objective" positions or from discovering historical Truth by improved methods or data, and thus requires we confront our ongoing and continually renewed complicities in narrating interconnected pasts and presents. As Trouillot says,

> The historicity of the human condition also requires that practices of power and domination be renewed. It is that renewal that should concern us most, even if in the name of our pasts. The so-called legacies of past horrors—slavery, colonialism, or the Holocaust—are possible only because of that renewal. And that renewal occurs only in the present. Thus, even in relation to The Past our authenticity resides in the struggles of our present. Only in that present can we be true or false to the past we choose to acknowledge. (Trouillot 1995, 151)

As an extension of this point, then, the dismissal and repeated dis-remembering of Native dispossession impacts those tribes still present today, whose own dispossession (partial or complete) also supports the settler colonial nation's gate of entry.

The installation of memory of the "extinct" tribes signals more than just a lament of the losses of colonialism. In fact, they *must* stand as more than just an observation of historic violence, population decimation, land struggles, and the unmaking of those tribes' geographies. Colonization does not rest in the past, which is precisely where it is continually relegated by the dominant society. This is why it is easy to remain silent on issues of colonization. Returning to Patrick Wolfe's articulation once again, Haozous presents us with an understanding of colonialism as process, a process that marches forward to maintain many of its spoils. The most valuable of these spoils, of course, is the land and the ongoing spatialities generated from the relationship with that land, including the nation-state itself. This memorialization thereby stands testimony to the ongoing struggles over indigenous geographies and colonial landscapes and serves as continuing witness to the process rather than to a concluded, past event.

At its current Santa Fe location, *Gate/Negate* has been placed near the main entryway to the capitol grounds, along the edge of the sidewalk and adjacent to the main parking lot, where one would be hard-pressed to ignore it. In presenting his narrative of immigration, nation, conquest, and colonial space, Haozous relies on a certain level of didacticism, even though or perhaps precisely because the extinguished tribal names and overall message confounds many viewers. This anticipated ignorance and naïveté leads Haozous to magnify his statements via sheer size, as well as through the durability of metal medium. In an interview about an explicitly environmental installation that Haozous placed at the Wheelwright Museum of the American Indian (also in Santa Fe), he confessed that "the American public is naïve. Bigger is better when there is a statement behind it" (McCloud 1995). This interest in scale seems a shared family ethic, as Haozous's father, Allan Houser, once explained that his increasingly larger sculptures fed an artistic purpose: "Working in a larger scale, I can express myself much better" (Hirchfelder n.d.).

Haozous, while certainly finding an aesthetic expression through scale, is also trying to eliminate silence and ignorance. He wrestles with obscure presentation and resistance to remembering forgotten memories by producing massive and enduring installations inviting encounter and interrogation. As Trouillot lamented elsewhere, silence now passes for innocence: "One now is innocent until proven guilty. Thus, claims of innocence can take the shape of silence" (Trouillot 2003, 28). By focusing on the work of the gate and the renewals of negation, Haozous calls out this silence. He dismisses such silence as being equal to innocence.

Afterword
Reclaiming Indigenous Geographies

Sacred Stone Camp, Standing Rock Indian Reservation,
North Dakota, April 1, 2016

Geographies in tension. Overlapped spatialities.

We Are Here to Protect. Mni Wiconi, Water Is Life. Why must the children run, demanding we ReZpect our Water?

Black Snake, Dakota Access Pipeline. Energy Transfer Partners. Pepper spray. Rubber bullets. Dogs. Arrests. Concussion grenades. Tazers. Armored vehicles. National Guard and "law" enforcement from far and wide have come to force a pipeline across the river. Water turned into a weapon, soaking bodies in freezing temperatures. Leave the roadway; leave private property.

What kind of geography will the US Army Corp engineer?

Keeper of the Sacred White Buffalo Calf Pipe, Arvol Looking Horse (Lakota/Dakota/Nakota) has an important message for us all. "In our prophecies," he says, "it is told that we are now at the crossroads: either unite spiritually as a global nation, or be faced with chaos, disasters, diseases, and tears from our relatives' eyes." Many listen and come. We need better transfers. To be better partners.

Looking Horse tells us that the police took the elders' "c'anupas (ceremonial pipes) and their prayer offerings. . . . They called our prayer sticks weapons."

While my initial interest in "Indian Villages" focused around non-Native appropriations of Indianness, the majority of this project turned toward understanding how Native communities and individuals have deployed indigenous spatialities, asserted the continuation and necessity of Native geographies. What I have tried to add to the conversation is explicit attention to the work of space and everyday acts of spatiality. The recent battle over the pipeline in North Dakota has brought these issues into stark relief for all to see, even if not all can or wish to see.

At Standing Rock, thousands of people gathered for nearly a year, coming together from across the globe to resist the threat posed by the Dakota pipeline not only to the Lakota and Dakota peoples, but to the water of the upper Missouri River, and to the land and nonhuman beings. Those gathered prayed and sang and communicated. They blocked construction equipment. They asked for assistance in protecting Native cultures and the rights of indigenous self-determination everywhere. During the confrontations at Standing Rock, Native people who were opposed to the placement of the Dakota Access Pipeline across that land likewise refused the label of "protestors." They insisted on recognition as Water Protectors. Rhetorically, this positioned the oil companies, law enforcement, politicians, and governmental agencies as harming water, and making choices to cause death. The slogan/prayer/stance was simple: "Water Is Life." The choice to protect water was not framed as a political choice, or as something that one might protest. The name Water Protectors signaled an ontological position that required consideration of a nonhuman world and reflected a cultural framework organizing the relationship between humans and water. The Water Protectors saw their presence in terms of a set of responsibilities that all human must uphold and that their spiritual teachings made explicit and unequivocal. Any other position meant opposition to life and thus a calculated sacrifice and designated death.

Anticolonial efforts inevitably press geographic and spatial tensions, but also have the potential to reveal the vast labor and resources needed to obscure and uphold such tensions. They offer a blueprint for redirecting that labor and channeling resources elsewhere. Reservations, allotment policies, removals, repatriation, sovereignty, gaming, and mascots. Pipelines.

When I began thinking about these issues, researching and writing about them, I felt quite alone. I was desperately trying to locate this work and to articulate what I saw as its potential contribution to the fields of comparative and critical ethnic studies, indigenous studies, and American studies. At the time I was not aware of the critical and cultural geography work being done, nor the work in the area of indigenous geography, since it was still just emerging and consolidating as a recognizable literature. I have been not only thrilled, but actually overwhelmed by the amount of scholarship that has developed in this area since that time. The research is still just developing, and I look forward to its trajectory. Perhaps more will hear Arvol Looking Horse's call.

It is with these thoughts in mind that I turn to some words about where this book has left me seeking more. The process of writing and thinking raised a number of questions that I either could not answer or did not address. Other questions arose in the fissures between my research interests and methods, or out of the limits of my capacity. So, I am now left with more questions about how indigenous space operates and its potential implications. One question posed to me by a colleague that I still cannot address is the relationship between indigenous language use and indigenous spatiality. In brief, is fluency in an indigenous language necessary for a truly indigenous geography to emerge, for a truly indigenous spatiality to be practiced? The phrase Mni Wiconi (Water Is Life) has clearly proven a powerful way to reframe or reclaim relations to that most fundamental element of existence for all (or at least most) beings on this planet. I have proceeded with the understanding that language is helpful, not crucial. But I cannot dismiss the way indigenous languages form the foundation of indigenous spatialities. I proceeded with an understanding that most of the examples I shared emerged from an English base, although this may not be accurate. In fact, multiple languages may be at work. Indigenous spatialities may also permeate through linguistic shifts. So, I am open to knowing more, as well as learning more about the implications of geographies generated through the valences of English, Spanish, or other languages of the colonizing cultures. I am certainly not prepared to determine those boundaries, and think they should be explored more carefully given the compelling arguments often made by folks like the late Darrell Robes Kipp about the fundamental relationship between language and indigenous cultures.

More generally I wonder about other mundane practices of spatiality, particularly indigenous spatiality, that need attention. If common sense drives the sustenance of spatialities, settler and indigenous alike, then it seems more work needs to fuse together cultural studies, geography, postcolonial theory, and indigenous studies. There is a good deal already being produced, but a solid consolidation of theories and practical examples would be most welcome. I have also been struck by the recent work outlining how indigenous mapping is inherently enfolded into the epistemologies and structures of modernity. I have been appreciative of the ways this growing work complicates, and issues much needed warnings about, the limits of such mapping as an anticolonial practice. These complications serve as good reminders of the flexibility of power and the incorporating capacity of capitalism and neoliberalism. Thus, my research leads me to want more guidance on how to think through these complications in productive ways and to frame the consequences of various spatial strategies for indigenous peoples (and for all peoples). In many ways, this likewise generates a desire to see more good examples where settler colonial spatialities have been successfully transformed to benefit indigenous peoples and geographies.

The Water Protectors have renewed my energies to know what this looks like, to more actively test the limits of possibility. If we can reframe how we connect all to water, can we also extend indigenous-centered responsibilities and relationships? Can we rescript and empower new spatialities that are not founded on and dependent on conquest?

Bibliography

Abbot, Lawrence. 1994. *I Stand in the Center of the Good: Interviews with Contemporary Native American Artists*. Lincoln: University of Nebraska Press.

Alderman, Derek H. 2000. "A Street Fit for a King: Naming Places and Commemoration in the American South." *Professional Geographer* 52 (4): 672.

———. 2003. "Street Names and the Scaling of Memory: The Politics of Commemorating Martin Luther King, Jr. within the African American Community." *Area* 35 (2): 163–173.

Alderman, Derek H., and Joshua Inwood. 2013. "Street Naming and the Politics of Belonging: Spatial Injustices in the Toponymic Commemoration of Martin Luther King Jr." *Social & Cultural Geography* 14 (2): 211–233.

Allen, John. 2003. *Lost Geographies of Power*. Malden, MA: Blackwell.

Alotta, Robert I. 1975. *Street Names of Philadelphia*. Philadelphia: Temple University Press.

Althusser, Louis. 2001. "Ideology and Ideological State Apparatus (Notes towards an Investigation)." In *Lenin and Philosophy, and Other Essays*. New York: Monthly Review Press.

Ambler, Marjane, Joan Hantz, Richard E. Little Bear, Patti Means, Mina Seminole, Linwood Tall Bull, Carol Ward, Bill Wertman, and Dave Wilson. 2008. *We, the Northern Cheyenne People: Our Land, Our History, Our Culture*. Lame Deer, MT: Chief Dull Knife College.

Anderson, Kat. 2013. *Tending the Wild: Native American Knowledge and the Management of California's Natural Resources*. Berkeley: University of California Press.

Anon. 1912. "Satanta's Name to a Town." *Haskell County Republican*, March 29.

———. 1990. "My Culture, My Art." *Colores*. http://video.pbs.org/video/1475048928/.

———. 2002. "Reconnections—Historic Council Grove -Al-Le-Ga-Wa-Ho Park." http://www.kawmission.org/places/kawmission/reconnectionshistoriccouncil-grove22.htm.

———. 2004. *Spirit of the Blackfeet*. https://vimeo.com/34051865.

———. 2011. "Brownback Announces First Five Notable Kansans." *Dodge City Daily Globe*, August 19.

———. 2012. "Satanta Day Ceremony Script." From author's collection.

———. 2013. "Chief and Princess Satanta." http://skyways.lib.ks.us/towns/Satanta/chief.html.

Ashcroft, Bill, Gareth Griffiths, and Helen Tiffin. 1998. *Key Concepts in Post-Colonial Studies*. London: Routledge.

Baker, Bill John. 2013. "Cherokee Nation, State Sign Historic Car Tag Compact." Cherokeephoenix.org, November 3. http://www.cherokeephoenix.org/Article/Index/7562.

Barker, Joanne. 2005. *Sovereignty Matters: Locations of Contestation and Possibility in Indigenous Struggles for Self-Determination*. Lincoln: University of Nebraska Press.

Barnd, Natchee Blu. 2008. "Inhabiting Indianness: US Colonialism and Indigenous Geographies." University of California, San Diego. http://www.escholarship.org/uc/item/7gc357ch.

———. 2010. "Inhabiting Indianness: Colonial Culs-de-Sac." *American Indian Culture & Research Journal* 34 (3): 27–45.

Basso, Keith H. 1996. *Wisdom Sits in Places: Landscape and Language among the Western Apache*. Albuquerque: University of New Mexico Press.

Bender, Margaret Clelland. 2002. *Signs of Cherokee Culture: Sequoyah's Syllabary in Eastern Cherokee Life*. Chapel Hill: University of North Carolina Press.

Benton Soil and Water Conservation District. 2013. "Blackberry Jones and the Invaders of Dunawi Creek." https://www.bentonswcd.org/assets/DunawiCrkTr10.pdf.

Ben-Zvi, Yael. 2007. "Where Did Red Go? Lewis Henry Morgan's Evolutionary Inheritance and U.S. Racial Imagination." *CR: The New Centennial Review* 7 (2): 201–229.

Berg, Lawrence D., and Jani Vuolteenaho. 2009. *Critical Toponymies: The Contested Politics of Place Naming*. Farnham, UK: Ashgate.

Bergland, Renée L. 2000. *The National Uncanny: Indian Ghosts and American Subjects*. Hanover, NH: Dartmouth College, University Press of New England.

Berkhofer, Robert F. 1978. *The White Man's Indian: Images of the American Indian from Columbus to the Present*. New York: Knopf (distributed by Random House).

Berlo, Janet Catherine, and Ruth B. Phillips. 1998. *Native North American Art*. Oxford: Oxford University Press.

Bertolini, Jim, and Janet Ore. 2012. "National Register of Historic Places Registration Form: Indian Memorial—Little Bighorn Battlefield National Monument." http://files.cfc.umt.edu/cesu/NPS/CSU/2011/11_12Ore_LIBI_National%20Register_Custer_Battlefield.pdf.

Bhabha, Homi K. 1990. *Nation and Narration*. London: Routledge.

————. 1994. *The Location of Culture*. London: Routledge.

Bierhorst, John. 1979. *A Cry from the Earth: Music of the North American Indians*. New York: Folkways Records. http://www.aspresolver.com/aspresolver.asp?GLMU;73029.

Biggs, Patricia. 2001. "No Longer 'All White Tukee': Minorities up 348% since 1990." *Arizona Republic*, January 21.

Bighead, Kate. 2004. "She Watched Custer's Last Battle, as Told to Thomas B. Marquis." In *The Custer Reader*, edited by P. A. Hutton. Norman: University of Oklahoma Press.

Billig, Michael. 1995. *Banal Nationalism*. London: Sage.

Biolsi, Thomas. 2005. "Imagined Geographies: Sovereignty, Indigenous Space, and American Indian Struggle." *American Ethnologist* 32 (2): 239–259.

Bird, S. Elizabeth, ed. 1996. *Dressing in Feathers: The Construction of the Indian in American Popular Culture*. 1st ed. Boulder, CO: Westview Press.

Blaut, James M. 1993. *The Colonizer's Model of the World: Geographical Diffusionism and Eurocentric History*. New York: Guilford Press.

Blee, Lisa, and Jean O'Brien. 2014. "What Is a Monument to Massasoit Doing in Kansas City? The Memory Work of Monuments and Place in Public Displays of History." *Ethnohistory* 61 (4).

Blomley, Nick, and Edgar Heap of Birds. 2004. "Artistic Displacements: An Interview with Edgar Heap of Birds." *Environment & Planning D: Society & Space* 22 (6): 799–807.

Blunt, Alison, and Cheryl McEwan. 2002. *Postcolonial Geographies*. London: Continuum.

Bohrer, Becky. 2003. "Little Bighorn Extends Honors to Indians." *Los Angeles Times*, June 22. http://www.freerepublic.com/focus/f-news/934246/posts.

————. 2005. "Northern Cheyenne Want to Change Names of Villages on Reservation." *Montana Standard*, September 7. http://mtstandard.com/news/state-and-regional /northern-cheyenne-want-to-change-names-of-villages-on-reservation/article _9a913bba-40f5-5419-9601-ea51f2b061d4.html.

Boyd, Maurice. 1981. *Kiowa Voices*. Vol. 1. Fort Worth: Texas Christian University Press.

Brooke, James. 1997. "Controversy over Memorial to Winners at Little Bighorn." *New York Times*, August 24.

Bruyneel, Kevin. 2007. *The Third Space of Sovereignty the Postcolonial Politics of U.S.-Indigenous Relations*. Minneapolis: University of Minnesota Press.

Bryan, Joe. 2009. "Where Would We Be without Them? Knowledge, Space and Power in Indigenous Politics." *Futures* 41 (1): 24–32.

Byrd, Jodi A. 2011. *The Transit of Empire: Indigenous Critiques of Colonialism*. Minneapolis: University of Minnesota Press.

Cajete, Gregory. 2000. *Native Science: Natural Laws of Interdependence*. Santa Fe, NM: Clear Light.

Carpio, Myla Vicenti. 2004. "Countering Colonization: Albuquerque Laguna Colony." *Wicazo Sa Review* 19 (2): 61–78.

Castree, Noel. 2004. "Differential Geographies: Place, Indigenous Rights and 'Local' Resources." *Political Geography* 23 (2): 133–167.

Césaire, Aimé. 1972. *Discourse on Colonialism*. New York: Monthly Review Press.

Chang. 1997. "A Meditation on Borders." In *Immigrants Out!: The New Nativism and the Anti-Immigrant Impulse in the United States*. Critical America series, edited by J. F. Perea. New York: New York University Press.

Chinuk Wawa Dictionary Project. 2012. *Chinuk Wawa: kakwa nsayka ulman-tilixam laska munk-kemteks nsayka = As our elders teach us to speak it*. Confederated Tribes of the Grand Ronde Community of Oregon. Seattle: University of Washington Press.

Coombes, Brad, Nicole Gombay, Jay T. Johnson, and Wendy S. Shaw. 2011. "The Challenges of and from Indigenous Geographies." In *A Companion to Social Geography*, edited by V. J. Del Casino, 472–489. Chichester, UK: Wiley-Blackwell.

Coulthard, Glen Sean. 2014. *Red Skin, White Masks: Rejecting the Colonial Politics of Recognition*. Minneapolis: University of Minnesota Press.

Craib, Raymond. 2004. *Cartographic Mexico: A History of State Fixations and Fugitive Landscapes*. Durham, NC: Duke University Press.

Cronon, William. 1983. *Changes in the Land: Indians, Colonists, and the Ecology of New England*. New York: Hill and Wang.

Davis, Gladys M. 1930. "And Thus Satanta Pushes On." *Satanta Chief*, June 12.

Deloria, Philip Joseph. 1998. *Playing Indian*. New Haven, CT: Yale University Press.

Deloria, Vine. 1969. *Custer Died for Your Sins: An Indian Manifesto*. New York: Macmillan.

———. 1994. *God Is Red: A Native View of Religion*. Golden, CO: Fulcrum.

Deloria, Vine, and American Indian Science and Engineering Society. 1991. *Indian Education in America: 8 Essays*. Boulder, CO: American Indian Science and Engineering Society.

Deloria, Vine, and Daniel R. Wildcat. 2001. *Power and Place: Indian Education in America*. Golden, CO: Fulcrum.

De Oliveira, Nicolas, Nicola Oxley, Michael Petry, and Michael Archer. 1994. *Installation Art*. Washington, DC: Smithsonian Institution Press.

Dilworth, Leah. 1996. *Imagining Indians in the Southwest: Persistent Visions of a Primitive Past*. Washington, DC: Smithsonian Institution Press.

Edensor, Tim. 2004. "Automobility and National Identity: Representation, Geography and Driving Practice." *Theory, Culture and Society* 21 (4-5).

Evans, Mei Mei. 2002. "'Nature' and Environmental Justice." In *The Environmental Justice Reader: Politics, Poetics, and Pedagogy*. Tucson: University of Arizona Press.

Ewers, John C., Jessie Wilber, Olga Ross Hannon, and Museum of the Rockies. 1976. *Blackfeet Indian Tipis: Design and Legend*. Bozeman, MT: Museum of the Rockies.

Farr, William E. 2009. *Julius Seyler and the Blackfeet: An Impressionist at Glacier National Park*. Norman: University of Oklahoma Press.

Fauntleroy, Gussie. 2011. "Looking Between the Lines." *Native Peoples Magazine* 24 (5): 30–35.

Ferguson, Thomas John. 1985. *A Zuni Atlas*. Norman: University of Oklahoma Press.

Flores, Richard R. 2009. "The Alamo: Myth, Public History, and the Politics of Inclusion." In *Contested Histories in Public Space: Memory, Race, and Nation*, edited by D. Walkowitz and M. Knauer. Durham, NC: Duke University Press.

Foote, A. Clifford. 2010. "News from the Desk of the Special Projects Coordinator A. Clifford Foote." http://nctribalhousing.org/special-projects-update/.

Freeman, Victoria. 2010. "'Toronto Has No history!' Indigeneity, Settler Colonialism, and Historical Memory in Canada's Largest City. (Report)." *Urban History Review* 38 (2): 21.

Fry, Timothy S. 1990. "The Unknown Indian Monument." *Heritage of the Great Plains* 23 (4).

Gease, Heidi Bell. 2003. "Former Hills Woman Helps Build Battle of Little Bighorn Memorial." *Rapid City Journal*, March 17.

Geerlings, Gerald K. 1983. *Wrought Iron in Architecture: An Illustrated Survey*. New York: Dover.

Goeman, Mishuana. 2013. *Mark My Words: Native Women Mapping Our Nations*. Minneapolis: University of Minnesota Press.

———. 2014. "Disrupting a Settler-Colonial Grammar of Place: The Visual Memoir of Hulleah Tsinhnahjinnie." In *Theorizing Native Studies*, edited by Audra Simpson and Andrea Smith. Durham, NC: Duke University Press.

Goldberg, David Theo. 1993. *Racist Culture: Philosophy and the Politics of Meaning*. London: Wiley-Blackwell.

Gordon, Alan. 2001. *Making Public Pasts: The Contested Terrain of Montréal's Public Memories, 1891–1930*. Montreal: McGill-Queen's University Press.

Gramsci, Antonio. 1971. *Selections from the Prison Notebooks of Antonio Gramsci*. New York: International Publishers.

Green, Rayna, and Massachusetts Arts and Humanities Foundation. 1975. "The Pocahontas Perplex: The Image of Indian Women in American Culture." *Massachusetts Review* 16 (4).

Haozous, Bob. 2000. "Gate/Negate: Artist's Statement." Handout from the capitol building, in author's collection.

———. 2007. "Indigenous Thoughts." *OpEdNews*. http://www.opednews.com/articles /opedne_bob_haoz_071208_indigenous_thoughts.htm.

Harjo, Joy. 2004. "Creation Story: The Jaune Quick-to-See Smith Survey." In *Postmodern Messenger, Jaune Quick-to-See Smith*. Tucson, AZ: Tucson Museum of Art.

Harley, John Brian. 2001. *The New Nature of Maps: Essays in the History of Cartography*. Baltimore, MD: Johns Hopkins University Press.

Harris, Cole. 2002. With cartography by Eric Leinberger. *Making Native Space: Colonialism, Resistance, and Reserves in British Columbia*. Vancouver: UBC press.

Hauʻofa, Epeli. 2008. *We Are the Ocean: Selected Works*. Honolulu: University of Hawaiʻi Press.

Heap of Birds, Edgar. 1987. "My Past, My People." In *Blasted Allegories: An Anthology of Writings by Contemporary Artists*, edited by B. Wallis. New York: New Museum of Contemporary Art; Cambridge, MA: MIT Press .

———. 1999. "From the Personal to the Political." In *Text and Image*. New York: The Photography Institute. http://www.thephotographyinstitute.org/www/1999/heap_of_birds.html.

———. 2007. "Artist Talk." In *Through Your Eyes: Walker Art Museum Project*. http://www.walkerart.org/archive/2/B22359FED4AC26DD615F.htm.

———. 2009. *Most Serene Republics*. Washington, DC: Smithsonian, National Museum of the American Indian.

Herman, RDK. 1999. "The Aloha State: Place Names and the Anti-Conquest of Hawaiʻi." *Annaassoamergeog Annals of the Association of American Geographers* 89 (1): 76–102.

Hertzberg, Hazel W. 1981. *The Search for an American Indian Identity: Modern Pan-Indian Movements*. 1st paper ed. Syracuse, NY: Syracuse University Press.

Hidalgo, Rianna. 2015. "Keeping His Tribe Alive." *Real Change*, February 18. http://realchangenews.org/2015/02/18/keeping-his-tribe-alive.

Hirchfelder, Arlene. n.d. "Allan Houser." *Artists and Craftspeople, American Indian Lives*. http://www.fofweb.com/History/MainPrintPage.asp?iPin=indoo18&DataType=Indian&WinType=Free.

Huhndorf, Shari M. 2001. *Going Native: Indians in the American Cultural Imagination*. Ithaca, NY: Cornell University Press.

Ingold, Tim. 2007. "Earth, Sky, Wind, and Weather." *Journal of the Royal Anthropological Institute* 13:S19–S38.

Jordan, Michael P. 2011. "Reclaiming the Past: Descendants' Organizations, Historical Consciousness, and Intellectual Property in Kiowa Society." PhD diss., University of Oklahoma.

Kansas Historical Society. 2014. "Satanta." *Kansapedia*. http://www.kshs.org/kansapedia/satanta/16881.

Kaplan, Thomas. 2010. "Britain Refuses Visas for Lacrosse Team." *New York Times*, July 15, p. 22.

Keller, Robert H., and Michael F. Turek. 1998. *American Indians and National Parks*. Tucson: University of Arizona Press.

Killsback, Leo. 2005. "Guest Opinion: Nationhood a Positive Change for Northern Cheyenne Tribe." *Billings Gazette*, September 16. http://billingsgazette.com/news

/opinion/guest-opinion-nationhood-a-positive-change-for-northern-cheyenne
-tribe/article_f6337c5c-c821-59bf-b918-da9f5bb08a68.html.

Kimmerer, Robin Wall. 2013. *Braiding Sweetgrass*. Minneapolis, MN: Milkweed Editions.

King, C. 2003. "De/Scribing Squ*w: Indigenous Women and Imperial Idioms in the United States." *American Indian Culture and Research Journal* 27 (2): 1–16.

Kipp, Darrell Robes. 2007. "Observations on a Tribal Language Revitalization Program." In *American Indian Nations: Yesterday, Today, and Tomorrow*, edited by G. Horse Capture, D. Champagne, and C. C. Jackson. Lanham, MD: AltaMira Press.

Korns, Kristan. 2012. "Hoopa Addresses Hard to Find." *Two Rivers Tribune*, September 11.

Last Soundtrack. 2007. "Bleeding Cowboys Font." Dafont.com. http://www.dafont.com/bleeding-cowboys.font.

Lefebvre, Henri. 1991. *The Production of Space*. Oxford, UK: Blackwell.

Leib, Jonathan. 2011. "Identity, Banal Nationalism, Contestation, and North American License Plates." *Geographical Review* 101 (1): 37–52.

Lippard, Lucy. 2004. "Jaune Quick-to-See Smith: Generosity with an Edge." In *Postmodern Messenger, Jaune Quick-to-See Smith*. Tucson, AZ: Tucson Museum of Art.

———. 2008. "Signs of Unrest: Activist Art by Edgar Heap of Birds." In *Most Serene Republics: Edgar Heap of Birds*, edited by K. E. Ash-Milby and T. Lowe. Washington, DC: National Museum of the American Indian, Smithsonian Institution.

Longfellow, Henry Wadsworth. 1942 [1855]. *The Song of Hiawatha*. Mount Vernon, NY: Peter Pauper Press.

Loomba, Ania. 2015. *Colonialism/Postcolonialism*. 3rd ed. Milton Park, UK: Routledge.

Lott, Eric. 1993. *Love and Theft: Blackface Minstrelsy and the American Working Class*. New York: Oxford University Press.

Lyons, Scott Richard. 2010. *X-Marks: Native Signatures of Assent*. Minneapolis: University of Minnesota Press. http://public.eblib.com/choice/publicfullrecord.aspx?p=548063.

Marshall, James. 1945. *Santa Fe: The Railroad That Built an Empire*. New York: Random House.

Massey, Doreen B. 2005. *For Space*. London: Sage.

McArthur, Lewis A. 2003. *Oregon Geographic Names*. 7th ed. Portland: Oregon Historical Society Press.

McCloud, Kathleen. 1995. "Site Haozous." *Santa Fe New Mexican*, September 29.

Meyer, Jonathan W. 1969. *A Transportation Study for Lame Deer, Montana*. Bozeman: Department of Civil Engineering and Engineering Mechanics, Montana State University.

Mignolo, Walter D. 1992. "Putting the Americas on the Map (Geography and the Colonization of Space)." *Colonial Latin American Review* 1 (1-2): 25–63.

Mills, Charles W. 1997. *The Racial Contract*. Ithaca, NY: Cornell University Press.

Mitchelson, Matthew L., Derek H. Alderman, and E. Jeffrey Popke. 2007. "Branded: The Economic Geographies of Streets Named in Honor of Reverend Dr. Martin Luther King, Jr." *Social Science Quarterly (Wiley-Blackwell)* 88 (1): 120–245.

Mithlo, Nancy Marie. 1998. "Lost O'Keefes/Modern Primitives: The Culture of Native American Art." In *Reservation X: The Power of Place in Aboriginal Contemporary Art*, edited by G. McMaster. Seattle: University of Washington Press; Hull: Canadian Museum of Civilization.

Moreton-Robinson, Aileen. 2015. *The White Possessive: Property, Power, and Indigenous Sovereignty*. Minneapolis: University of Minnesota Press.

Moses, L. G. 1999. *Wild West Shows and the Images of American Indians, 1883–1933*. Albuquerque: University of New Mexico Press.

Musqueam Indian Band. 2006. *Musqueam: A Living Culture*. Victoria, BC: CopperMoon Communications.

———. 2007. "Musqueam Community Profile: Knowing Our Past, Exploring Our Future." http://www.edo.ca/downloads/musqueam-community-profile.pdf.

Nash, Catherine. 2002. "Cultural Geography: Postcolonial Cultural Geographies." *Progress in Human Geography* 26 (2).

Nkrumah, Kwame. 1966. *Neo-Colonialism: The Last Stage of Imperialism*. New York: International Publishers.

Northern Cheyenne Cultural Commission. 2007. "Draft of A Resolution of the Northern Cheyenne Tribal Council Expanding Ordinance No. 10 (97)." http://www .cheyennenation.com/executive/cheyenne_language_resolution.pdf.

Ohnesorge, Karen. 2008. "Uneasy Terrain: Image, Text, Landscape, and Contemporary Indigenous Artists in the United States." *American Indian Quarterly* 32 (1).

Olp, Susan. 2014. "Indian Memorial at Little Bighorn Completed." *Casper Star-Tribune*, June 9. http://trib.com/news/state-and-regional/indian-memorial-at-little-bighorn -completed/article_8d287c68-84e4-5afd-89fa-93720d188c53.html.

Olsen, Daniel M., and Henk van Assen. 2009. *Ranch Gates of the Southwest*. San Antonio, TX: Trinity University Press.

Pappan, Chris. 2015. Personal communication. September 3, 2015.

Pfeuffer, Charyn. n.d. "Artist Keeps Blackfeet Tradition Alive." *OnTrak*. http://www .ontrakmag.com/guardipee.

Pierotti, Raymond, and Daniel R. Wildcat. 2002. "Being Native to This Place." In *American Indians in American history, 1870–2001: A Companion Reader*, edited by S. Evans. Westport, CT: Praeger.

Powell, Peter J. 1979. *People of the Sacred Mountain: A History of the Northern Cheyenne Chiefs and Warrior Societies, 1830–1879 : With an Epilogue, 1969–1974*. San Francisco: Harper and Row.

Pratt, Mary Louise. 1992. *Imperial Eyes: Travel Writing and Transculturation*. London: Routledge.

Radcliffe, Sarah A. 2011. "Third Space, Abstract Space and Coloniality: National Subaltern Cartography in Ecuador." In *Postcolonial Spaces: The Politics of Place in Contemporary Culture*, edited by A. Teverson and S. Upstone, 129–145. Basingstoke, UK: Palgrave Macmillan.

———. 2017. "Geography and Indigeneity I: Indigeneity, Coloniality and Knowledge." *Progress in Human Geography* 41 (2). http://journals.sagepub.com/doi/abs/10.1177/0309132515612952?journalCode=phgb.

Rader, Dean. 2011. *Engaged Resistance: American Indian Art, Literature, and Film from Alcatraz to the NMAI*. Austin: University of Texas Press.

Rafael, Vicente L. 2000. *White Love: And Other Events in Filipino History*. Durham, NC: Duke University Press.

Raibmon, Paige. 2008. "Unmaking Native Space: A Genealogy of Indian Policy, Settler Practice, and the Microtechniques of Dispossession." In *The Power of Promises: Rethinking Indian Treaties in the Pacific Northwest*, edited by A. Harmon. Seattle: Center for the Study of the Pacific Northwest, in association with University of Washington.

Reardon, Jenny, and Kim Tallbear. 2012. "'Your DNA Is Our History': Genomics, Anthropology, and the Construction of Whiteness as Property." *Current Anthropology* 53 (Supp. 5): S233–S245.

Rifkin, Mark. 2013. "Settler Common Sense." *Settler Colonial Studies* 3 (3-4): 322–340.

———. 2014. *Settler Common Sense: Queerness and Everyday Colonialism in the American Renaissance*. Minneapolis: University of Minnesota Press.

Robinson, Charles M. 1998. *Satanta: The Life and Death of a War Chief*. Austin, TX: State House Press.

Romm, Jeff. 2002. "Coincidental Order of Environmental Injustice." In *Justice and Natural Resources: Concepts, Strategies, and Applications*. Washington, DC: Island Press.

Rose-Redwood, Reuben. 2009. "Indexing the Great Ledger of the Community: Urban House Numbering, City Directories, and the Production of Spatial Legibility." In *Critical Toponymies: The Contested Politics of Place Naming*, edited by L. D. Berg and J. Vuolteenaho. Farnham, UK: Ashgate.

Rushing, W. Jackson. 1999. *Native American Art in the Twentieth Century*. London: Routledge.

———. 2005. "'In Our Language': The Art of Hachivi Edgar Heap of Birds." *Third Text* 19 (4): 365–384.

Said, Edward W. 1979. *Orientalism*. New York: Vintage.

———. 1993. *Culture and Imperialism*. 1st ed. New York: Knopf (distributed by Random House).

Sasse, Julie, and Jaune Quick-to-See Smith. 2004. *Postmodern Messenger, Jaune Quick-to-See Smith.* Tucson, AZ: Tucson Museum of Art.

Schultz, James Willard. 1926. *Signposts of Adventure; Glacier National Park as the Indians Know It.* Boston: Houghton Mifflin.

Senogles, Renee. 1996. Interview by Dagmar Seely. *People of the Seventh Fire: Returning Lifeways of Native America.* Ithaca, NY: Akwe:kon Press.

Simpson, Audra. 2014. *Mohawk Interruptus: Political Life across the Borders of Settler States.* Durham, NC: Duke University Press.

Slocum, Elizabeth. 2007. "'Insurgent Messages': Hachivi Edgar Heap of Birds and Building Minnesota." *Museo* 7 (Spring).

Snyders, Tom, and Jennifer O'Rourke. 2001. *Namely Vancouver: The Hidden History of Vancouver Place Names.* Vancouver: Arsenal Pulp Press. http://site.ebrary.com/id/10209258.

Spence, Mark David. 1999. *Dispossessing the Wilderness: Indian Removal and the Making of the National Parks.* New York: Oxford University Press.

Spivak, Gayatri Chakravorty. 1987. *In Other Worlds: Essays in Cultural Politics.* New York: Methuen.

Stanley, Francis. 1968. *Satanta and the Kiowas.* Berger, TX: Jim Hess Press.

Stark, Heidi Kiiwetinepinesiik. 2012. "Marked by Fire: Anishinaabe Articulations of Nationhood in Treaty Making with the United States and Canada." *American Indian Quarterly* 36 (2): 119–149.

Strong, Pauline Turner. 2013. *American Indians and the American Imaginary: Cultural Representation across the Centuries.* Boulder, CO: Routledge.

Tribal Council of the Northern Cheyenne Tribe. 2010. "A Resolution of the Northern Cheyenne Tribal Council Officially Approving the Official License Plate Design of the Northern Cheyenne Tribe." http://www.cheyennenation.com/license.html.

Trouillot, Michel-Rolph. 1995. *Silencing the Past: Power and the Production of History.* Boston, MA: Beacon Press.

——. 2003. *Global Transformations: Anthropology and the Modern World.* New York: Palgrave Macmillan.

United States Bureau of Indian Affairs. 1964. "A Preliminary Urban Planning Program for Lame Deer, Montana." Washington, DC: United States Department of the Interior, Bureau of Indian Affairs.

United States. Indian Claims Commission. 1979. *United States Indian Claims Commission, August 13, 1946–September 30, 1978: Final Report.* Washington: The Commission.

Vancouver City Council. 2014. "Protocol to Acknowledge First Nations Unceded Traditional Territory." City council minutes. June 24, Motion on Notice B-3. http://council.vancouver.ca/20140624/documents/motionb3.pdf.

Veracini, Lorenzo. 2014. "Understanding Colonialism and Settler Colonialism as Distinct Formations." *Interventions* 16 (5): 615–633.

Vizenor, Gerald Robert. 1994. *Manifest Manners: Postindian Warriors of Survivance*. Hanover, NH: Wesleyan University Press.

Vogel, Virgil J. 1991. "Placenames from Longfellow's 'Song of Hiawatha.'" *Names* 39 (3).

Warhus, Mark. 1998. *Another America: Native American Maps and the History of Our Land*. 1st St. Martin's Griffin ed. New York: St. Martin's Griffin.

Warrior, Robert Allen. 1995. *Tribal Secrets: Recovering American Indian Intellectual Traditions*. Minneapolis: University of Minnesota Press.

Washburn, Betty. 2010. Personal Correspondence with the author.

———. n.d. "Scalp Dance Notes." By Betty Washburn, in author's collection.

Washoe Cultural Preservation Office. n.d. "Wa She Shu: 'The Washoe People' Past and Present." https://www.fs.usda.gov/Internet/FSE_DOCUMENTS/stelprdb5251066 .pdf.

Willdorf, Nina. 2000. "Art Loses Its Edge." *Chronicle of Higher Education*, June 9.

Wolfe, Patrick. 2006. "Settler Colonialism and the Elimination of the Native." *Journal of Genocide Research* 8 (4): 387–409.

Wood, Debora. 1998. "Art and Transformation." *Issues in Integrative Studies* 16.

Worster, Donald. 1979. *Dust Bowl: The Southern Plains in the 1930s*. New York: Oxford University Press.

Index